# Development of Hand Skills in Children

***Editors:***

*Jane Case-Smith, EdD, OTR*

*and*

*Charlane Pehoski, MA, OTR, FAOTA*

**Disclaimers**

"This publication is designed to provide accurate and authoritative information in regard to the subject matter covered. It is sold or distributed with the understanding that the publisher is not engaged in rendering legal, accounting, or other professional service. If legal advice or other expert assistance is required, the services of a competent professional person should be sought."

—From the Declaration of Principles jointly adopted by the American Bar Association and a Committee of Publishers and Associations.

ISBN:0-910317-78-X

It is the objective of the American Occupational Therapy Association to be a forum for the free expression and interchange of ideas. The opinions and positions expressed by the contributors to this work are their own and not necessarily those of either the editors or the American Occupational Therapy Association.



PRINTED IN THE UNITED STATES OF AMERICA

# TABLE OF CONTENTS

**Copyright Information and Disclaimers** ..................................................... ii
**About the Authors** ..................................................... v
**Foreword** ..................................................... vii

Chapter 1: Central Nervous System Control of Precision Movements of the Hand ..................... 1
*by Charlane Pehoski, MA, OTR, FAOTA*

Chapter 2: Eye–Hand Coordination ..................................................... 13
*by Rhoda P. Erhardt, MS, OTR, FAOTA*

Chapter 3: In-Hand Manipulation Skills ..................................................... 35
*by Charlotte E. Exner, PhD, OTR, FAOTA*

Chapter 4: Therapeutic Fine-Motor Activities for Preschoolers ..................................................... 47
*by Carol Anne Myers, MS, OTR/L*

Chapter 5: Handwriting: Evaluation and Intervention in School Settings ..................................... 63
*by Susan J. Cunningham Amundson, MS, OTR/L*

Chapter 6: Developing Scissors Skills in Young Children ..................................................... 79
*by Colleen Schneck, ScD, OTR, and Carmela Battaglia, MS, OTR/L*

Chapter 7: Neurodevelopmental Treatment for the Young Child with Cerebral Palsy ................ 91
*by Elizabeth Danella, MOT, OTR, FAOTA, and Laura Vogtle, MS, OTR/L*

Chapter 8: Upper-Extremity Casting: Adjunct Treatment for the Child with Cerebral Palsy ....... 111
*by Audrey Yasukawa, MOT, OTR/L*

# About the Authors

***Charlane Pehoski, MA, OTR, FAOTA*** has a bachelor's degree in occupational therapy from the College of St. Catherine and a master's degree in occupational therapy from Boston University. She also has a certificate in physical therapy from the same institution. Pehoski is certified both in the administration of the Sensory Integration and Praxis Tests and in neurodevelopmental treatment. Presently, she is the Director of Occupational Therapy and the Director of Client and Family Services at the University-Affiliated Program at the Shriver Center in Waltham, Massachusetts. She is also in the process of completing her doctoral work at Boston University. Her thesis research is on the development of in-hand manipulation skills in preschool children. Pehoski is a Fellow of the American Occupational Therapy Association.

***Rhoda P. Erhardt, MS, OTR, FAOTA,*** received her bachelor of science degree in occupational therapy from the University of Illinois and her master's degree in child development and family relations from North Dakota State University. She was certified in pediatric neurodevelopmental treatment in London, England. The former director of the Easter Seal Mobile Therapy Unit in Fargo, North Dakota, she is currently in private practice, providing education and consultation services to a variety of health agencies and educational systems in the Midwest, as well as presenting workshops worldwide. She has served on the editorial board of the *American Journal of Occupational Therapy* and on the Board of Occupational Therapy Practice of the state of North Dakota, and was enrolled in the AOTA Roster of Fellows in 1983. The use of her developmental prehension assessment for neurologically impaired children has been described in published journal articles, videotapes, and her first book, *Developmental Hand Dysfunction*. Her videotape on normal hand development received an award from the American Academy for Cerebral Palsy and Developmental Medicine. Her new developmental visual assessment (revised 1989) and series of videotapes are correlated with her second book, *Developmental Vision Dysfunction*, published in 1990.

***Charlotte E. Exner, PhD, OTR, FAOTA,*** is Assistant Professor and Chairperson of Occupational Therapy at Towson State University, Towson, Maryland. She has a PhD in Human Development from the University of Maryland (1991). Her master of science degree in education of the severe and profoundly handicapped is from Johns Hopkins University (1980) and her bachelor of science degree in occupational therapy is from Ohio State University (1974). Her research has been in the areas of splinting with children with cerebral palsy, tactile functions in normal children and those with cerebral palsy, development of in-hand manipulation skills in normal preschool children, and instrument development in the latter two areas. Her dissertation addressed assessment of the Test of In-Hand Manipulation by Exner.

***Carol Anne Myers, MS, OTR/L,*** has a bachelor's degree in music therapy from the University of Georgia, a master of science degree in occupational therapy from Boston University, certification to administer the Sensory Integration and Praxis Tests from Sensory Integration International, and is licensed by the Commonwealth of Massachusetts. She worked as a music therapist in a psychiatric hospital for 4 years before returning to school for her training in occupational therapy. For the past 9 years she has worked as an occupational therapist with special-needs preschoolers, the last 6 of these years with the Brookline-Newton Early Childhood Collaborative. She is a consultant to local nursery schools and is a frequent speaker at workshops for therapists and teachers.

***Susan J. Cunningham Amundson , MS, OTR/L,*** currently is a lecturer at and coordinator of the Rural Pediatric Training Project in the Division of Occupational Therapy, Department of Rehabilitation Medicine, University of Washington in Seattle. For 8 years, she has worked in the public schools of Washington and Alaska. Her presentations regarding children's handwriting dysfunction have been heard by national and international audiences.

***Colleen Schneck, ScD, OTR,*** was Senior Occupational Therapist at The Harry S. Tack Education Center, Sewickley, Pennsylvania, and Adjunct Assistant Professor of Occupational Therapy at the University of Pittsburgh. Since receiving her bachelor's degree in occupational therapy in 1977, Dr. Schneck has worked with infants, preschoolers, and school-aged children in a variety of hospital and school-based programs.

***Carmela Battaglia, MS, OTR/L,*** is Assistant Professor of Occupational Therapy at the University of Pittsburgh. Since receiving her bachelor's degree in occupational therapy in 1971, Battaglia has worked with preschoolers, school-aged children, and adolescents in both acute care and rehabilitation facilities and in a variety of school-based programs. She is currently a doctoral student in the Graduate School of Public Health at the University of Pittsburgh.

***Elizabeth Danella, MOT, OTR, FAOTA,*** obtained her occupational therapy education at Tufts University and received an advanced master's degree in pediatrics from the University of Florida. She taught in neurodevelopmental treatment certification courses as an occupational therapy instructor and continues to present workshops on the NDT approach to handling the young child with cerebral palsy. She presently is the director of the Pediatric Re/Habilitation Department at the Children's Hospital of New Jersey and specializes in the treatment of newborns and young children with motor disorders.

***Laura Vogtle, MS, OTR/L,*** has 20 years of occupational therapy experience with pediatrics. Her major interest has been children with central nervous system disorders. She has been a neurodevelopmental treatment occupational therapy instructor for 10 years, with extensive teaching experience in neurodevelopmental therapy and a variety of other topics as well. This includes teaching in the graduate school of special education at the University of Virginia. Another strong interest is research design and critique. At present, Vogtle is a doctoral candidate in educational evaluation at the University of Virginia, having received a master's degree in educational research in May 1991. She has been involved with a number of research efforts related to technology applications.

***Audrey Yasukawa, MOT, OTR/L,*** is chief of occupational therapy at La Rabida Children's Hospital and Research Center in Chicago. She has worked extensively as a pediatric occupational therapist. Previously she taught physically handicapped children for the Chicago school system. She is certified in the neurodevelopmental treatment approach to children with cerebral palsy, and adult hemiplegics. Nationally, Yasukawa has taught numerous upper-extremity casting workshops and spoken at several AOTA annual conferences. She also has published articles on casting of the upper extremity. Audrey received her bachelor's degree in special education for the physically handicapped from Northern Illinois University and her master's in occupational therapy from Western Michigan University.

# Foreword

The hand is an exquisite instrument that enables us to perform skills that are uniquely human. Our daily occupations depend on the dextrous skill of our hands. As occupational therapists, we often focus on developing hand skills to improve a child's independence in self-care, play, and school activities. In the past, our literature has focused on grasp and reach as the primary elements of fine-motor skills. Recent studies have demonstrated that the development of hand skills encompasses more than the refinement of grasp and release. In acknowledging that the continuum of hand skill development extends well beyond the mastery of reach and prehension, the authors of this volume have explored the complex processes of manipulation and handling of objects and tools.

The development of skillful hand use as required in our daily occupations is influenced by a myriad of variables. This volume's authors discuss factors such as cognition, vision, and somatosensory systems that influence hand skills in the context of child development. They also explain the therapeutic approaches and strategies that enhance the development of hand skills and the child's ability to play, manipulate, and skillfully use tools.

The authors give us a perspective on the complexity of hand skill development, provide frameworks for understanding manipulation and tool use, and then discuss occupational therapy approaches for improving the child's hand skills.

In the first chapter, Pehoski explains neurophysiological research as it relates to precision hand skill. Recent research has demonstrated that hand movements are directly controlled by corticospinal pathways that are separate from the midbrain and brainstem-level pathways that control posture and mobility. Since most hand musculature is innervated through neuro pathways that originate in the motor cortex, conscious attention and thought are required to direct the hand's movements. Therefore manipulation, while tactually and visually guided, must be "learned" and has a strong relationship to the individual's cognitive development. Pehoski discusses the implications of this direct relationship between the hand's movements and the motor cortex to occupational therapy and to the child's skill development.

In the second chapter, Erhardt provides a detailed description of the development of eye–hand coordination based on over 30 years of research. The development of visual and hand skills and the development of cooperative action between these two systems are essential to the child's ability to play, perform self-care, and achieve in school. Erhardt discusses the literature on eye–hand coordination from an occupational therapist's perspective and develops a framework for intervention based on the neurophysiological and developmental understandings we have of eye–hand coordination. Her model provides a framework for matching activities to the child's cognitive level, motivation, visual skill, somatosensory perception, and eye–hand interaction.

In chapter three, Exner explains her classification system for in-hand manipulation skills. These advanced skills may be observed in coin or key manipulation or in pen or pencil use and represent the highest levels in continuum of fine-motor-skill development. Exner also describes her recently developed test of in-hand manipulation. She relates the specific manipulation skills to activities in which they are used and to the musculature that is involved in each skill. The complexities of these skills and their interrelationships are discussed based on validity studies of the Test of In-Hand Manipulation by Exner.

The fourth chapter defines and illustrates best practice principles in therapy with preschoolers who have fine-motor delays. Myers uses biomechanical and neurodevelopmental models to select and create activities to enhance the preschoolers' development of hand skills. Play activities and manipulatives are adapted to meet the specific needs of preschoolers with immature hand skills. Explicit descriptions of therapeutic activities include how they should be performed and why the activities effectively promote hand skills.

In chapter five, handwriting evaluation and intervention are described. Handwriting represents a major occupation of the school-aged child, and school-based occupational therapists are often viewed as critical resources for children who have handwriting problems. Cunningham Amundson describes a framework for analyzing handwriting problems and describes five therapeutic approaches that can be used singly or in combination to prevent or improve handwriting problems. The neuromuscular, biomechanical, multisensory, motivational, and acquisition approaches are used to evaluate handwriting and select appropriate therapeutic activities for educational settings. Understanding the underlying skills of handwriting and the factors involved in learning to write legibly is essential to designing an effective therapy approach. Cunningham Amundson advocates flexible application of complementary therapeutic approaches as the child's developmental needs change and the academic requirements for this skill increase.

In chapter six, Schneck and Battaglia explain the development of scissors skills. Scissors are one of the first tools that a child learns to use. Yet a number of basic skills are needed for the child to accomplish efficient scissors use. Cutting involves the use of both hands in separate movements and separate planes as well as isolated, coordinated, intrinsic movements of the hand manipulating the scissors. Due to the demanding complexity of this tool activity, occupational therapists begin with therapeutic activities that practice the components of scissors skill. In addition to describing intervention to enhance scissors use, Schneck and Battaglia describe methods and equipment to adapt this task for children with fine-motor delays.

In chapter seven, Danella and Vogtle explain the neurodevelopmental treatment (NDT) approach to developing hand skills in young children with cerebral palsy. After a review of the problems observed in cerebral palsy and a discussion of neurodevelopmental treatment principles, the authors provide a model and examples of intervention. They emphasize the importance of postural control, sensation, spine and joint alignment, qualitative elements of movement, and functional outcomes. The NDT framework includes analysis of postural-motor components and simultaneous task analysis. By analyzing the environment, task, and the child's skills, therapy can be directed toward the specific motor components needed in activities that are appropriate for and selected by the child.

In chapter eight, Yasukawa discusses casting techniques for the upper extremity that may be used with children who have cerebral palsy. Casting procedures and precautions are explained. The various types of commonly used upper-extremity casts are identified and criteria for selection of each type are provided.

This book melds the research and theories of occupational therapy and other disciplines into models of practice for use with children who have fine-motor delays. Our concepts about hand skills are expanded from knowledge of basic reach, grasp, and manipulation skills. These concepts are critical given that almost all daily living occupations depend on the individual's ability to manipulate objects and use tools. This book contributes to our understanding of how hand skills develop in children and of how occupational therapy can promote these essential life skills in children.

## Acknowledgments

The editors thank Ann Henderson, PhD, OTR, for her inspiration and leadership in pediatric occupational therapy. She has helped us understand the complexities of hand skill development and has promoted our abilities as therapists to facilitate the child's acquisition of competent hand function.

The editors also thank the members of the AOTA Developmental Disabilities Special Interest Section Standing Committee, 1989–1993, for their direction in developing this book and for reviewing the chapters:

- Debra Cook, MS, OTR
- David Nelson, PhD, OTR
- Kathy Stewart, MS, OTR
- Barbara Burris Wavrek, MHS, OTR

*—Jane Case-Smith*
*DDSIS Standing Committee*

# 1

# CENTRAL NERVOUS SYSTEM CONTROL OF PRECISION MOVEMENTS OF THE HAND

*Charlane Pehoski*

## INTRODUCTION

The development of precision hand skills is essential to the child's ability to play, use tools,and perform self-care skills and is often the focus of occupational therapy. When a child has a motor deficit, such as poor control of refined hand skills, knowledge of how the brain controls these movements can be very helpful in planning the therapeutic program. This chapter reviews some of the neurophysiology literature pertinent to the control of skilled hand use. Much of this research involves monkeys, and although there is an obvious gap between the abilities of monkeys and those of humans, this information is extremely helpful for a beginning understanding of how the brain controls skilled movements. The final section of this chapter summarizes therapy principles that emerge through application of this research to human function and development.

The development of the hand as an instrument of skill is considered an important factor in man's evolutionary history. This history involves both structural changes in the hands themselves as well as changes in the central nervous system (CNS). Compared to man's ancestors and the hands of the present-day great apes, modern man's fingers are straighter and the distal phalanx broader, thus allowing a larger sensory pad at the end of the fingers, and there is an increase in the length of the thumb and index finger making finger tip opposition possible (Napier, 1962).

As the structural changes in the hand allowed for its evolution as an instrument capable of versatility, changes also evolved in the CNS to support these functions. The cortical mantel expanded and the overall size of the corticospinal tract increased. This tract is believed to provide the ability to fractionate movement, that is, to perform movements involving the independent use of separate muscles. This is particularly representative of finger movements (Lawrence & Kuypers, 1968a).

One purpose of this chapter is to review the literature that supports two premises important for understanding CNS control over precision movements of the hand. One premise is that there is not one motor system that develops proximal to distal, but rather two systems, one that controls distal movements and the other that controls postural and more proximal movements. For control of the upper extremity, this means that manipulation and reach are dependent on two different groups of descending fiber systems. Skilled manipulation is dependent on the corticospinal track while reach is primarily dependent on pathways originating from the midbrain and brainstem (Lemon, 1990). The second premise is that skilled hand function is dependent on sensory feedback and that this feedback might be different for the hand musculature than for the muscles of the trunk and shoulder girdle.

The ability to use the hands with skill not only has had a long evolutionary course, but also has a long developmental course. At birth, man's hands are crude instruments. Grasp is reflexive, triggered by tactile or proprioceptive input. Not until 8 to 9 months of age is the child able to produce a neat precision grip using the finger tips and the thumb (Bayley, 1969). But true manipulation of objects, exemplified by the coordinated movement of the distal fingers in activities such as buttoning, may not be efficient until around 4 years of age (Folio & Fewell, 1983; Stutsman, 1948). Several studies indicate that these changes may be related to the maturation of neural structuring supporting skilled hand use. These studies will also be reviewed. Finally, therapeutic implications from these studies will be discussed.

## Two Motor Systems

Kuypers and his colleagues have done a great deal to increase our knowledge about the influence of the descending motor pathways' control over body movements (Kuypers, 1981). This research involves two main areas. One looks at the termination of the descending motor tracts onto interneurons and motoneurons in the spinal cord (Kuypers, 1960; Kuypers, Fleming, & Farinholt, 1962), and the second selectively lesions these tracts in monkeys so the residual behaviors can be observed (Lawrence & Kuypers, 1968 a,b). From this research, two fundamentally different motor systems are described, one arising from brainstem structures, which is felt to be related to posture and proximal control, and the other arising from cortical structures, which is felt to be related to distal control.

### Organization of Descending Motor Tracts in the Spinal Cord

The only way control can be exerted over the muscles of the body is through synapses onto motoneurons in the ventral gray area in the spinal cord (Lawrence & Kuypers, 1968a, b). Descending motor tracts achieve most of their connections indirectly through an interneuron pool in the gray area of the spinal cord rather than synapsing directly onto the motoneuron pool itself. One important exception is corticospinal fibers that make direct connections to motoneurons, particularly motoneurons hand musculature (Kuypers, 1960; Kuypers, 1962). These connections are thought to provide the ability to fractionate movements as exemplified by control over individual finger movements (Lawrence & Kuypers, 1968a).

Both the interneuron pool and the motoneuron pools in the spinal cord are somatotopically organized—that is, the neurons to the distal extremities musculature are separate from those to the proximal/trunk musculature. It was by looking at descending motor tract input to these two areas that led Kuypers (1963) to suggest two fundamentally different motor systems—a brainstem system, which can again be divided into two parts, and a corticospinal system. Lawrence and Kuypers (1968a, b) felt that these systems represented increasingly greater control over the refinement of movement.

The Group A, or medial brainstem pathways, contain information from the reticular formation, vestibular complex, superior colliculus, and intersitual nucleus of Cajal. They primarily synapse on interneurons for the trunk and proximal muscles of the extremities. The authors suggest that these pathways represent the basic system by which the brain controls movement. These Group A pathways are felt to be specifically related to the maintenance of erect posture, integration of movements of the body and the limbs, movements of the body and head, and with direction of the course of progression. The Group B, or lateral brainstem system, to which the rubrospinal tract is a major contributor, supplements the control of the Group A pathways and provides for independent flexor-biased movements of the extremities, particularly of the distal parts (Kuypers, 1981). Lastly, the corticospinal tracts provide precision and speed, and especially in primates, the ability to isolate individual movements as exemplified by finger movements (Kuypers, 1981). Of particular importance to this discussion are the direct corticospinal connections onto motoneurons of the distal extremities—that is, neurons that do not synapse first onto interneurons but connect directly to the motoneurons themselves. As mentioned, the ability to perform the independent finger movements necessary for skilled manipulation, including the ability to perform a neat pincer grasp, is felt to be mediated through these connections.

**Figure 1. Schematic diagram of (lateral) corticospinal pathways from the primary motor cortex to the spinal cord. The diagram indicates the direct corticospinal fibers felt to be responsible for the ability to fractionate movement and those that synapse on interneurons.**

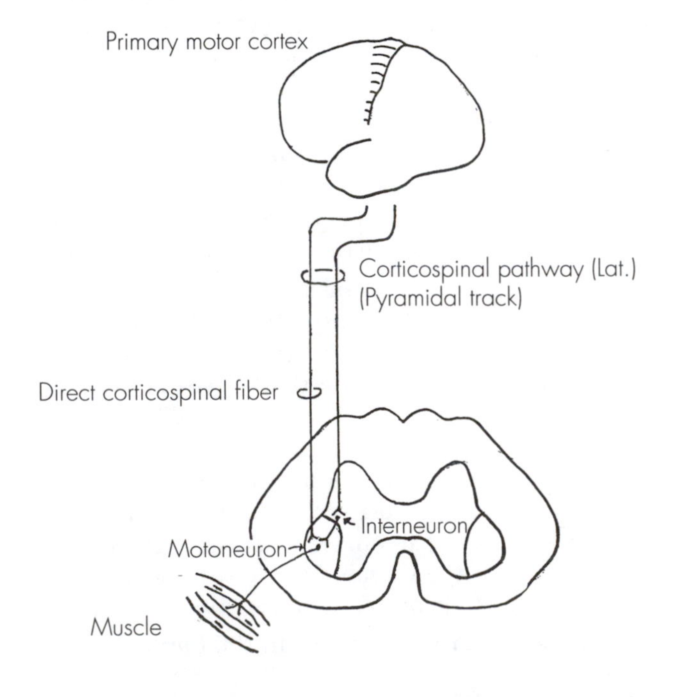

## Behavioral Characteristics of Lesioned Monkeys

Based on these findings, it is interesting to note the behavioral characteristics of animals who have undergone selective lesions of these different descending motor systems. Lawrence and Kuypers (1968a) demonstrated that monkeys with bilateral pyramidotomies interrupting the corticospinal fibers can immediately sit, walk, run, and climb postoperatively. Yet, even after 3 to 4 years of survival, these animals could only close the hand by flexing all the fingers together. They were never able to demonstrate a precision grip using the thumb and index finger and had difficulty with the release of food from the hand even though they had no difficulty releasing the bars of the cage when climbing.

On the other hand, if animals were subject to lesions of the Group A or medial group of descending brainstem motor tracts, they were noted to sit with the limbs and trunk flexed, the head slumped and the shoulders elevated. They stood in a crouched position and walked unsteadily. When they reached for food, they demonstrated a coarse proximal ataxia but no detectable impairment in the agility and speed of hand movements (Lawrence & Kuypers, 1968b). That is, despite the proximal disturbance in movement and posture the animals could isolate individual fingers to prehend small pieces of food.

## Necessity of Corticospinal and Motor Cortex Input to Hand Skill

It therefore appears that the motor system can be functionally divided into two components, one controlling axial and postural movements and the other controlling skilled movements, particularly movements of the hands. This latter activity is provided by input from the motor cortex itself through the corticospinal tract, *particular fibers that make direct connection with the hands*. As an example, direct corticospinal connections are only seen in animals who possess the ability to manipulate objects with the fingers. This is true even when the shape of the animal's hand would accommodate this function. Lawrence and Hopkins (1976) point out that the shape of the hand and fingers of the squirrel monkey is almost identical to that of the rhesus monkey but in the squirrel monkey there are no direct cortical motor connections. This animal can only grasp small objects by using total finger flexion. The rhesus monkey has direct cortical connections and the ability to manipulate objects.

Apparently, in monkeys (and probably in man), there does not appear to be any tracts within the central nervous system that can adequately substitute for the function of the corticospinal or pyramidal tract. To study this issue, Lawrence and Hopkins (1976) performed bilateral pyramidotomies on five infant rhesus monkeys. The operations were performed from between 5 days to 4 weeks after birth. In addition, two infant monkeys without pyramidal lesions were studied and served as developmental controls for the experimental animals. The authors note that the early development of reaching and hand use did not differ between the two groups of animals, but that later, the control animals studied began to exhibit a large range of small movements of the hand and forearm, including a pincer grasp. Kuypers (1962, 1981) had earlier demonstrated that the bulk of the direct corticospinal connections are not present in the newborn monkey but develop gradually over the first 6 months largely in parallel to the development of a pincer grasp. A neat pincer grasp was not seen in the experimental animals. That is, at the time of development when the corticospinal fibers would conceivably influence the control of the hand, the experimental animals continued to pick up small objects with the fingers closing as a unit. The experimental animals also demonstrated difficulty in releasing objects so that food had to be removed from the closed hand by the mouth. These behaviors continued for the 3 years of the animals' survival. The authors concluded that in the absence of the corticospinal or pyramidal tract in the monkey, no other descending pathway appears to be able to establish the input necessary for the development of relatively refined finger movements.

It is not only transection of the pyramidal tract that can cause loss of skilled hand function. A precision grip using the thumb and index finger to pick up a small object also fails to develop when the contralateral primary motor cortex is lesioned in infant monkeys. This is not surprising since at least 40% of pyramidal fibers originate from the primary motor cortex (Wise & Evarts, 1981). Passingham, Perry, and Wilkinson (1978) attempted to look at the possibility that a lesion of the primary motor cortex in one hemisphere might allow the undamaged hemisphere to substitute for the missing information. They placed unilateral lesions in either the primary motor cortex or the primary motor cortex and the primary sensory cortex in infant monkeys. None of the lesioned animals developed the ability to pick up objects using a neat precision grip. The authors therefore concluded that neither the intact contralateral hemisphere nor any subcortical areas in monkeys can take over the direction of a precision grip if the primary motor cortex or the pyramidal tracts are lesioned.

## Role of the Primary Motor Cortex

The importance of the primary motor cortex to precision activity of the hand was further demonstrated in a study by Muir and Lemon (1983). The authors trained a monkey to squeeze two levers either by using a precision grip or by squeezing a rubber cylinder using a power grip. They then recorded from the primary motor cortex where neurons with direct connections to the intrinsic muscles of the hand had been isolated. Discharge from these neurons was monitored as the animal performed the two grip patterns. In addition, electromyograph (EMG) responses from the hand muscles were also monitored. What the authors found was a small population of neurons in the primary motor cortex that were principally active when the animal was asked to make discrete movements such as those required for a precision grip. This was true despite the fact that the EMG information indicated that some muscles were active for both types of movements. The authors suggested that in a precision grip the muscles are used in a fractionated manner to position the fingers independently, while in the power grip the monkey produces a general cocontraction of all the digits. They therefore concluded that primary motor cortex neurons have a particular role during the fractionation of muscles required for discrete movements such as the movements of individual fingers. These neurons are not nearly as active when the animal moves the fingers as a unit as demonstrated in a power grip.

Muir (1985) demonstrated that in man, and to a lesser extent in monkeys, there is variability in the temporal pattern of activation of hand muscles during a precision grip. That is, when the EMG activity of various hand muscles is monitored, the muscles used in a precision grip fired at different times during the grip of the object. In contrast, during a power grip, the muscles tend to be coactivated. Muir suggests that, "corticomotoneuronal neurons undoubtedly play an important part in the selection and sequencing of muscles during precision grip movements" (p. 171).

One of the factors that is striking about the motor maps of the cortex is that movements involving distal muscles are represented by a larger cortical area than those representing more proximal motions. As Cheney (1985) has pointed out, this is true despite the smaller size of the distal muscles. He indicates that this disparity indicates a preferential role of motor cortex in distal movement. Other authors have indicated that this relationship may be particularly strong for the intrinsic muscles of the hand. As an example, in man, corticospinal connections are said to be particularly heavy to motoneurons, innervating the intrinsic muscles of the hand (Schoen, 1964). Clough, Kernell, and Phillips (1968) further delineated these findings by studying cortical excitation in monkeys on specific hand and forearm muscles. They found that the greatest response to cortical stimulation was in the intrinsic muscles of the hand and in extensor digitorum communis. This is not surprising since the intrinsic hand muscles are particularly important to precision handling. Long, Conrad, Hall, and Furler (1970) indicated that the ability to rotate an object in the fingertips requires the interossei muscles of the hand as well as muscles of the thenar eminence. The extrinsic muscles are said to provide "gross motion" and compression force (Long et al., 1970). Fahrer (1981) suggests that the extrinsic muscles provide the force and the intrinsic muscles the precision. These findings are consistent with the clinical picture of the lesioned animals discussed earlier, animals who were unable to produce a precision grip and who had difficulty with the release of objects (Lawrence & Hopkins, 1976; Lawrence & Kuypers, 1968a). The latter activity requires both the intrinsic muscles and the extensor digitorum communis.

It has been suggested that the primary motor cortex is also related to the initiation of willed as opposed to instinctual motor responses. As an example, Hoffman and Luschell (1980) found that most of the motor cortex neurons they studied failed to show a change in discharge to natural chewing movements in monkeys, yet these same neurons did respond when the monkey was trained to produce chewing movements on demand. This difference was seen despite the fact that similar patterns of EMG muscle activity were observed in both tasks. Evarts (1981) suggests that this study implies the greater concern of the motor cortex with learned as opposed to instinctual movements. Cortically induced movements tend to involve long-term training (Asanuma & Arissian, 1984).

## Role of Sensory Input

### Closed Loop Feedback Control Systems

When an animal initiates a learned or willed movement, at least two modes of control can be used. One mode is to operate open loop. This generally involves ballistic or fast movements. In this mode, the command for the movement is said to be preprogrammed prior to movement onset, and, once the movement is launched, it is not appreciably changed by sensory stimuli. The second mode of control is said to operate closed loop. Closed loop control occurs with small, precise movements (Fromm & Evarts, 1977). In this mode, the system does respond to sensory feedback and is felt to be dependent on this feedback (Desmond & Godaux, 1978; Fromm & Evarts). Evarts (1981) suggests that

> Closed loop feedback control systems (both segmental and suprasegmental) have evolved to operate under conditions where errors are small. When errors are large, centrally programmed movements involving open loop systems generate large control pulses that in turn activate intense muscle activity that is not under reflex feedback control. (p. 1103)

To operate closed loop (e.g., to use sensory feedback in the control of movement), the motor cortex must have access to sensory information from the periphery. This input originates from two sources: directly from the dorsal columns and indirectly through cortico-cortical connections with the sensory cortex itself (Asanuma & Arissian, 1984). When a movement occurs, sensory feedback generated by the movement feeds back to the efferent columns in the primary motor cortex that activate the contracting muscles themselves. This input may be quantitatively different for the distal and proximal muscle. It has been suggested that in the monkey motor cortex, cutaneous modalities are better represented in the hand area, where deep receptors (muscle, joints, tendons) are better represented in the proximal limbs (Lemon, 1981). In addition, several authors have found that motor cortex neurons in the monkey hand area activated by cutaneous stimuli have receptive fields almost exclusively on the palmar surface of the hand and fingers (Lemon; Rosen & Asanuma, 1972; Strick & Preston, 1982). Lemon found that these neurons were particularly active when a monkey performed small exploratory movements without the aid of vision (e.g., seeking a sunflower seed concealed in the examiner's hand).

Evarts (1981) suggests that the motor cortex has a special role in "controlling movements under guidance by somatosensory inputs" (p. 1085). Patients who have lost sensory input to the extremities, either through deafferentation due to a severe peripheral sensory neuropathy (Rothwell, Traub, Day, Obeso, Thomas, & Marsden, 1982) or a lesion confined to the primary sensory cortex (Jeannerod, Michel, & Prablanc, 1984), experience significant curtailment of everyday hand functions. Activities such as crumpling a piece of paper in the hand or buttoning are extremely difficult or impossible for these patients. Rothwell et al. (1982) suggest that the difficulty of their patient "lay in the absence of any automatic reflex correction in his voluntary movements and also to sustain constant levels of muscle contraction without visual feedback" (p. 515).

## Sensory Motor Mechanisms in the Hand

As mentioned, sensory information is important to the control of precision movements. Evarts (1981) has suggested that sensory information, such as that from the muscle spindle receptors, is meant to help correct for small error. Tracy (1980) indicates that one role of the spindle receptors is to help brace the joint and then to allow a controlled movement to proceed by adjusting the strength of the opposing contraction. The intrinsic muscles of the hand are among the muscles of the body most heavily innervated with muscle spindle receptors (Cooper, 1960). Devandan, Ghosh, and John (1983) state that the spindle receptors in the hand muscles contribute to skilled use through three functions: coordinating the forces between intrinsic and extrinsic muscles, rapidly correcting for errors, and providing the stability for the highly mobile digits.

Some of these stretch receptors appear to be facilitated by other sensory input. Marsden, Merton, and Morton (1971) demonstrated that when flexion of the distal thumb was perturbed (brief stretch to the long thumb flexor), an increase in muscle output was recorded. If the thumb was anesthetized, eliminating sensory input but not interfering with the muscle spindle receptors located in the forearm, this reflexive response was eliminated. The authors, therefore, suggest that this servo action in the thumb is apparently dependent on peripheral sensation. Of particular interest is that when this same experiment was carried out on the big toe, anesthetizing the toe did not abolish the response (Marsden, Morton, & Merton, 1977). This was also true when the experiment was carried out on the infraspinatus and the pectoralis major muscles. Marsden, et al. (1977) suggest that servo action in the big toe, infraspinatus, and pectoralis major may be based predominantly on muscle receptor alone, and that the servo action of the thumb is dependent on additional peripheral sensory input. They state that this is "presumably connected with the fact that the thumb executes fine manipulative and exploratory movements in which somatic and muscular afferents have to work in harness" (p. 534).

Therefore, not only are the motor pathways different for postural/axial muscles and distal muscles, but sensory input concerned with these movements also appears to demonstrate differences. The two motor systems have evolved to support very different functions and it is therefore not surprising to find differences in the sensory mechanisms controlling these functions.

## Role of Sensory Motor Mechanisms in the Hand

Several factors are necessary if the hand is to manipulate an object with skill. One is that the fingers move independently of each other. As an example, holding and rotating an object held in the fingertips would be extremely difficult to accomplish if the thumb and fingers work as a unit. As discussed, nature has provided direct corticospinal fibers that allow these independent movements to occur. A smooth, effortless performance would also be seriously curtailed if the hand did not have adequate sensory information about the movement of the object in the hand.

Johansson and Westling (1984) found that when sensitive measurements were taken of an adult using a pincer grasp to lift a small object, the force controlling the grip and the force used to provide the lift of the object were developed in parallel. That is, adults do not *grip* and then *lift*, rather these two forces increase together so that when the object leaves the surface the grip is *just* sufficient to maintain the object in the fingers. This small margin of safety in the grip force varied with objects of different textures (e.g., the force would be greater for an object covered with silk, and less for one covered with sandpaper). The change in force for different textures is felt to be dependent on cutaneous input. It is eliminated if the fingers are anesthetized (Westling & Johansson, 1984). Tactile input is also necessary to update the grip force so it can adjust to any small slippage of the object as it is held in the hand.

The functional significance of this ability is obvious. If the grip force is too tight, further manipulation of the object will be impossible. If it is too loose, the object will be dropped. Tactile input is necessary to maintain this balance. In fact, when the subjects in the Johansson and Westling (1987) study were asked to voluntarily open the fingers to let the object drop, they often reported that it felt like their fingers were adhering to the object. In this instance, the tactile input generated by the "slip" of the object was triggering an automatic motor response that helped to maintain the grip and prevent the object from slipping.

## Maturational Changes in Structures Supporting Precision Hand Skills

Humans are not born with the ability to use the hand with skill. This is a function that develops over time and is most probably dependent on the maturation of structures supporting precision handling. As mentioned, in monkeys the direct corticospinal connections to motoneurons of the distal extremities develop gradually over the first 6 months of life (Kuypers, 1962), and this development parallels the emergence of a neat pincer grasp (Kuypers, 1981; Lawrence & Hopkins, 1976). Yet, in humans, myelinization in the pyramidal tract (the tract in which the corticospinal fibers travel) is not completed until about 3 years of age (Yakoviev & Lecours, 1967). There is also reason to suspect that further maturation in this pathway continues until at least 11 years of age.

Using electromagnetic stimulation of the motor cortex and monitoring EMG responses in the abductor digiti minimi of the hand, Koh and Eyre (1988) tested subjects from 33 weeks of gestational age to 50 years of age. They found that the conduction velocities in the corticospinal paths increased with age (e.g., information traveled faster) and did not reach adult levels until 11 years of age. In addition, they found that cortical stimulation could not be detected in the resting muscle of children under 8 years of age. The original research design for this study was to record EMG activity following cortical stimulation in a resting muscle. This worked for the older children and adults, but children aged 8 years and younger had to perform an isometric contraction before the increased EMG activity following the stimulation could be seen. The authors state that, "this need for facilitation suggests that the cortical interneurons or pyramidal neurons have a higher threshold for activation in young children than in adults or that the alpha motor neurons are less readily excited by the activity evoked in the descending motor pathways after electromagnetic stimulation" (p. 1351).

Using a procedure similar to that used to test the pincer grasp of adults (Johansson & Westling,1984). Forssberg, Eliasson, Kinoshita, Johansson, and Westling (1991) explored the parameters inherent in lifting an object using a pincer grasp in children, 8 to 10 months to 13 to 14 years of age. As mentioned previously, adults demonstrate a smooth, parallel gain in the grip force and the lift force (Johansson & Westling), but in children under the age of 2 years, these two forces tend to be generated independently. That is, for those under the age of 2 years, most of the grip force was in place before the lift force started. The younger children also tended to push down during the grip phase and to exhibit a greater grip force than adults. Grip force was significantly greater for children 5 years or younger, particularly those 4 years and younger, than in adults. After 8 to 10 years of age, the performance of the children in this study did not significantly differ from the adults.

As mentioned earlier, if the grip force on an object is too great, manipulation of an object by the fingers is impossible or, at best, limited. Apparently a child's ability to hold an object in a pincer grasp with just sufficient force so that the object does not slip, forces similar to those used by an adult, does not appear until about 4 years of age or older. Therefore, it might be expected that skills dependent on precision handling would be difficult or inefficient for children under 4 years. Tables 1 and 2 indicate the time of acquisition of two developmental tasks: buttoning, a task requiring manipulation by the fingers, and sequential thumb-to-finger opposition, a task that demonstrates control over individual fingers. As indicated by these studies (Folio & Fewell, 1983; Stutsman, 1948), buttoning seems to be possible for the 3-year-old children, but it does not appear to be efficient until about 4 years of age.

A manipulative activity such as buttoning also requires the ability to isolate the movements of the individual fingers. Tables 1 and 2 seem to suggest that the ability to fractionate individual fingers, as demonstrated by rapid sequential finger to thumb opposition, is also not efficient until about 4 years of age.

**Table 1. Percentage of Children Passing or Time in Seconds for Three Items on the *Merrill Palmar Mental Scales* (Stutsman, 1948)**

| | Age in Months | | | | | | |
|---|---|---|---|---|---|---|---|
| | 24–29 | 30–35 | 36–41 | 42–47 | 48–53 | 54–59 | 60–65 |
| Button one button (% passing) | 19% | 72% | | | | | |
| Button two buttons (seconds) | | 170 sec. | 50 sec. | 34 sec. | 30 sec. | 23 sec. | 19 sec. |
| Thumb/finger opposition (% passing) | 0% | 6% | 35% | 50% | | | |

**Table 2. Percentage of Children Passing Two Items From the *Peabody Developmental Motor Scales* (Folio & Fewell, 1983)**

| | Age in Months | | | | | |
|---|---|---|---|---|---|---|
| | 24–29 | 30–35 | 36–41 | 42–47 | 48–59 | 60–71 |
| Button two buttons in 20 seconds (% passing) | 0% | 2% | 11% | 38% | 65% | 76% |
| Thumb/finger opposition in 8 seconds (% passing) | 0% | 0% | 4% | 22% | 72% | 82% |

Denckla (1973, 1974) studied finger-thumb opposition in older children. In these studies, the time for completion of 20 sequential, thumb-finger patterns was recorded. The author found a marked decrease in the time needed to perform this task in 5-, 6-, and 7-year-old children (Denckla, 1973), but no significant difference among 9,10, and 11 years of age (Denckla, 1974). It is interesting to note that these changes with age appear similar to the changes in age found in the conduction velocity of the corticospinal tracts reported by Koh and Eyre (1988). That is, differences tended to diminish after 8 to 11 years of age.

Taken together, these studies might suggest that efficient fingertip manipulation begins at about 4 years of age and possibly improves until about 8 to 11 years of age. This can only be a tentative hypothesis since so few studies exist that describe how hand skills, particularly manipulative skills, develop. Other areas of upper extremity development, such as reach (Bower, Broughton, & Moore, 1970; Von Hofsten, 1979, 1982; Von Hofsten & Lindhagen, 1976) and grasp (Halverson, 1931; Holstein, 1982; Von Hofsten & Ronnquist, 1988), have been explored, but only a few studies *specifically* examine how children manipulate objects. These studies have been with infants where manipulation is crude and primarily confined to wrist rotation, transferring the object hand to hand, and poking or fingering (Rochat, 1989; Ruff, 1982, 1984). This lack of research is surprising since reach and grasp are only the beginning action on an object. True hand skill involves manipulation with the fingers.

## THERAPEUTIC IMPLICATIONS

The research studies explored in this chapter have several implications for the therapeutic program addressed to the child with poor hand skills. This information would appear to be particularly applicable to the child who is having difficulty with refined hand movements.

*1. The difference in the control of proximal and distal musculature should be reflected in therapeutic programs.*

For a long time, therapists have looked at the proximal-distal progression in the maturation of upper extremity function and assumed a *causal* relationship between shoulder stability and hand function. There is no doubt that an indirect biomechanical relationship exists between the shoulder and the hand. Increased shoulder stability provides an improved base of support for controlled hand movements, particularly for hand function in space when the hand is not on a supporting surface (e.g., reaching out to put a peg in a Light Bright). But in many instances when a precision task is to be completed with the hands, the arm is externally stabilized, either against the trunk (as in threading a needle) or against a table or other working surface. In the child with poor shoulder stability, appropriate external support to the shoulder and arm can often be provided so that manipulative tasks can be introduced prior to the optimal development of proximal stability.

In addition, the neural mechanisms controlling precision movements of the hand and those controlling shoulder stability are different (Lawrence & Kuypers, 1968 a, b; Lemon, 1990). For example, the ability to hold a pencil in a mature tripod grasp and to perform the small, fractionated movements necessary for efficient pencil control requires mechanisms separate from those that provide stability of the shoulder. Therefore, a therapeutic program directed to these two areas should not be looked at as a sequential process but as a simultaneous process. Improving hand skill should not wait until shoulder stability is maximized.

Furthermore, the small, individual movements of the fingers involved in manipulation rely on different neuronal mechanisms than those involved in a mass grasp pattern or a power grip (Muir & Lemon, 1983). Therapeutic activities such as swinging from a trapeze or activities that require resisted mass grasp can strengthen a child's wrist and upper extremities and help improve the postural stability of the hand. However, the development of manipulative skill will almost always require practice with activities that elicit more refined use of the hand. In particular, attention may need to be directed to the function of the intrinsic muscles of the hand since they are critical to skilled use.

*2. Cortically controlled movements require the attention and active participation of the subject.*

Evarts (1981) suggested that cortical control over movements is more related to learned as opposed to instinctual movements. Cortical activities tend to be conscious, self-directed, and superimposed on more automatic background activity. Postural responses are automatic. They can be elicited without the subject's conscious awareness. In fact, an activity such as balance in sitting is best elicited if the subject is *not* trying to consciously control his or her balance. On the other hand, cortically directed activities require a shift of attention to the activity. Therefore the activities need to be sufficiently stimulating and challenging to promote learning or to evoke a learned response. Often the art of working with children is finding an activity at the interest and skill level of the child, that engages refined movements of the hand, and that elicits the child's enthusiastic and full cooperation.

*3. Practice is necessary for the development of skilled hand use.*

Asanuma and Arissian (1984) indicate that cortically induced movements often require long periods of training. This therapist has found that the play history of a child experiencing fine-motor difficulties often indicates that the amount of time the child spends in fine-motor tasks is much less than that of the child's peers. As an example, the nursery school teacher might report that the child avoids the "fine-motor table" and instead prefers to play with the large blocks. Parents may report that the child dislikes or does not play with small toys such as Legos. Therefore, the child who appears to need *more* experience in fine-motor work is actually receiving *less* input than his or her peers. Again, it is often the role of the occupational therapist to provide activities that not only address the specific area of fine-motor weakness, but activities that will engage and maintain the child's attention. Such activities should appeal to the child's interests and result in a feeling of mastery (see Meyers, chapter 4).

*4. Training can often increase the skill level of the hand despite damage to the corticospinal or pyramidal tract.*

It has been suggested that the motor cortex (Passingham et al., 1978) and the corticospinal, pyramidal tract (Lawrence & Hopkins, 1976) are necessary for the execution of discrete movements of the finger. Yet, several studies using monkeys have indicated that training helped the return of more functional use of the hand despite loss of these structures (Growden, Chambers, & Liu, 1967; Lawrence & Hopkins, 1976; Schwartzman, 1978). When training was not provided, the animals continued to demonstrate lack of use of the involved extremity. As an example, Lawrence and Kuypers (1968) did not find any major change in the ability of their animals to produce a precision grip after the initial period of recovery following surgery. Chapman and Wiesendanger (1982) suggest that this was due to the fact that no systematic effort was made to retrain use of the affected hand. Lesioned animals are known to use the hand spontaneously in activities such as climbing and swinging from the bars of the cage, but to ignore its use for fine movements of the hand. This marked neglect improved with training through activities that motivated the animal to use the extremity. Training results were particularly successful when the animal had sustained an incomplete lesion of the pyramidal tract (Chapman & Wiesendanger; Lawrence & Hopkins, 1976). As an example, in Lawrence and Hopkins' (1976) study of pyramidal tract lesions in infant monkeys, the authors state that

> some degree of RIFM [relatively independent finger movements] developed in the animal with incomplete pyramidal lesions. They appeared early and devel-

oped to the greatest degree in the hand contralateral to the largest number of spared fibers *but became apparent only during and after the training period in the hand opposite to the tract containing very few spared fibers*. [Emphasis added] (p. 248)

Therefore it would appear that intervention that motivated the animals to use the hand was necessary for the animals' recovery. The environment alone and activities such as using the hand in climbing were not sufficient to improve the animals' precision use of the hand. This same need for intervention appears to be true in humans.

*5. More studies describing the changes in hand function with changes in age are needed.*

It is very hard to determine when certain hand skills should be introduced to a child. As indicated, 4 years of age may be a time when efficient manipulation is beginning (Folio & Fewell, 1983; Stutsman, 1948). Yet, little literature exists that describes how infants progress from holding an object in a mass grasp to the ability to manipulate an object in the fingertips (see Exner, chapter 3).

The development of hand skills has traditionally been an area of specialization for occupational therapists. It is an area both physiologically and psychologically related to the basic tenets of our profession: concern for functional, goal-directed activities. Occupational therapists are therefore particularly well suited to make a significant contributions to research in this area.

## References


Asanuma, H., & Arissian, S. (1984). Direct and indirect sensory input pathways to the motor cortex; its structure and function in relation to learning of motor skills. *Journal of Physiology, 39,* 1-19.

Bayley, N. (1969). *Bayley scales of infant development.* New York: Psychological Corporation.

Bower, T.G.R., Broughton, J.M., & Moore, M.K. (1970). Demonstration of intention in the reaching behavior of neonate humans. *Nature, 228,* 679-681.

Brinkman, J., & Kuypers, H.G. (1973). Cerebral control of contralateral and ipsilateral arm and finger movements in the split-brain rhesus monkey. *Brain, 96,* 653-674.

Chapman, C., & Wiesendanger, M. (1982). Recovery of function following unilateral lesions of the bulbar pyramid in the monkey. *Electroencephalography and Clinical Neurophysiology, 53,* 374-387.

Cheney, P.D. (1985). Role of cerebral cortex in voluntary movements: A review. *Physical Therapy, 65,* 624-635.

Clough, J.F.M., Kernell, D., & Phillips, C.G. (1968). The distribution of monosynaptic excitation from the pyramidal tract and from primary spindle afferents to motoneurons of the baboon's hand and forearm. *Journal of Physiology, 198,* 145-166.

Cooper, S. (1960). Muscle spindle and other muscle receptors. In G.H. Bourne (Ed.), *The structure and function of muscle.* New York: Academic Press.

Denckla, M.B. (1973). Development of speed in repetitive and successive finder movements in normal children. *Developmental Medicine and Child Neurology, 15,* 635-645.

Denckla, M.B. (1974). Development of motor coordination in normal children. *Developmental Medicine and Child Neurology, 16,* 729-741.

Devandan, M.S., Ghosh, S., & John, K.T. (1983). A quantitative study of muscle spindle and tendon organs in the Bonnet monkey. *The Anatomical Record, 207,* 265-266.

Evarts, E.V. (1981). Role of motor cortex in voluntary movements in primates. In J.M. Brookhart and V.B. Mountcastle (Eds.), *Handbook of physiology, section I, volume II, motor control, part 2.* Bethesda, MD: American Physiological Society.

Fahrer, M. (1981). Interdependent and independent actions of the fingers. In R. Tubiana (Ed.), *The hand.* Philadelphia: Saunders.

Folio, M.R., & Fewell, R.R. (1983). *Peabody developmental motor scales.* Texas: DLM Teaching Resources.

Forssberg, H., Eliasson, A.C., Kinoshita, H., Johansson, R.S., & Westling, G. (1991). Development of human precision grip I: Basic coordination of force. *Brain Research, 85,* 451-457.

Fromm, C., & Evarts, E. (1977). Relation of motor cortex neurons to precisely controlled and ballistic movements. *Neuroscience Letters, 5,* 259-265.

Garnett, R., & Stephens, J.A. (1981). Changes in the recruitment threshold of motor units produced by cutaneous stimulation in man. *Journal of Physiology, 311,* 463-473.

Growden, J.H., Chambers, W.W., & Liu, C.N. (1967). An experimental student of cerebellar dyskinesia in the rhesus monkey. *Brain, 90,* 603-630.

Halverson, H.M. (1931). An experimental study of prehension in infants by means of systematic cinema records. *Genetic Psychological Monograph, 10,* 107-285.

Hoffman, D.S., & Luschell, E.S. (1980). Precentral cortical cells during a controlled jaw bite task. *Journal of Neurophysiology, 44,* 333-348.

Holstein, R.R. (1982). Development of prehension in normal infants. *American Journal of Occupational Therapy, 36,* 170-176.

Issler, H., & Stephens, J.A. (1983). The maturation of cutaneous reflexes studied in the upper limb in man. *Journal of Physiology, 335,* 643-654,

Jeannerod, M., Michel, F., & Prablanc, C. (1984). The control of hand movements in a case of hemianesthesia following a parietal lesion. *Brain, 107,* 899-920.

Jenner, J.R., & Stephens, J.A. (1982). Cutaneous reflex responses and their central nervous pathways studied in man. *Journal of Physiology (London), 333,* 405-419.

Johansson, R.S., & Westling, G. (1984). Role of glabrous skin receptors and sensorimotor memory in automatic control of precision grip when lifting rough or more slippery objects. *Experimental Brain Research, 56,* 550-564.

Johansson, R.S., & Westling, G. (1987). Signals in tactile afferents from the fingers eliciting adaptive motor responses during precision grip. *Experimental Brain Research, 66,* 141-154.

Koh, T.H., & Eyre, J.A. (1988). Maturation of corticospinal tracts assessed by electromagnetic stimulation of the motor cortex. *Archives of Diseases of Children, 63,* 1347-1352.

Kuypers, H.G., (1960). Central cortical projections to motor and somatosensory cell groups. *Brain, 83,* 167-184.

Kuypers, H.G. (1962). Corticospinal connections postnatal development in the rhesus monkey. *Science, 138,* 678-680.

Kuypers, H.G. (1963). The organization of the motor system. *Journal of Neurology, 4,* 78-91.

Kuypers, H.G. (1981). Anatomy of the descending pathways. In J.M. Brookhart and V.B. Mountcastle (Eds.), *Handbook of physiology, section I, volume II, motor control, part I.* Bethesda, MD: American Physiological Society.

Kuypers, H.G., Fleming, W.R., & Farinholt, J. (1962). Subcorticospinal projections in the rhesus monkey. *Journal of Comparative Neurology, 118,* 107-137.

Lawrence, D.G., & Hopkins, D.A. (1976). The development of motor control in the rhesus monkey: Evidence concerning the role of corticomotoneuronal connections. *Brain, 99,* 235-254.

Lawrence, D.G., & Kuypers, H.G. (1968a). The functional organization of the motor system in monkey. I. The effects of bilateral pyramidal lesions. *Brain, 91,* 1-14.

Lawrence, D.G., & Kuypers, H.G. (1968b). The functional organization of the motor system in monkey. II. The effect of lesions of the descending brain-stem pathways. *Brain, 91,* 15-36.

Lemon, R.N. (1981). Functional properties of monkey motor cortex neurons receiving afferent input from the hand and fingers. *Journal of Physiology, 31,* 497-519.

Lemon, R.N. (1990). Contributions to the history of psychology: LXVII Henricus (Hans) Kuypers F.R.S. 1925-1989. *Perceptual and Motor Skills, 70,* 1283-1288.

Long, C., Conrad, P.W., Hall, E.A., & Furler, M.S. (1970). Intrinsic-extrinsic muscle control of the hand in power grip and precision handling. *The Journal of Bone and Joint Surgery, 52A,* 853-867.

Marsden, C.D., Merton, P.A., & Morton, H.B. (1971). Servo action and stretch reflex in human muscle and its apparent dependence on peripheral sensation. *Journal of Physiology, 216,* 21p-22p.

Marsden, C.D., Merton, P.A., & Morton, H.B. (1977). The sensory mechanism of servo action in human muscle. *Journal of Physiology, 265,* 521-535.

Muir, R.B. (1985). Small hand muscles in precision grip: A corticospinal prerogative. *Experimental Brain Research, 10,* 155-173.

Muir, R.B., & Lemon, R.N. (1983). Corticospinal neurons with a special role in precision grip. *Brain Research, 261,* 312-316.

Napier, J.R. (1962). The evolution of the hand. *Scientific America, 207,* 56-62.

Passingham, R., Perry, H., & Wilkinson, F. (1978). Failure to develop a precision grip in monkeys with unilateral neocortical lesions made in infancy. *Brain Research, 145,* 410-414.

Rochat, P. (1989). Object manipulation and exploration in 2 to 5 month old infants. *Developmental Psychology, 25,* 871-884.

Rothwell, J.C., Traub, M.M., Day, B.L., Obeso, J.A., Thomas, P.K., & Marsden, C.D. (1982). Manual motor performance in a deafferented man. *Brain, 105,* 515-542.

Rosen, I., & Asanuma, H. (1972). Peripheral afferent input to the forelimb area of the monkey cortex: Input-output relations. *Experimental Brain Research, 14,* 257-273.

Ruff, H. (1984). Infants' manipulative exploration of objects: Effect of age and object characteristics. *Developmental Psychology, 20,* 9-20.

Ruff, H. (1982). Role of manipulation in infant's responses to invariant properties of objects. *Developmental Psychology, 18,* 682-691.

Schoen, J.H.R. (1964). Comparative aspects of the descending fiber system in the spinal cord. In J.C. Eccles and J.P. Schade (Eds.), *Progress in brain research. Organization of the spinal cord. Vol. II.* Amsterdam: Elsevier.

Schwartzman, R.J. (1978). A behavioral analysis of complete unilateral section of the pyramidal tract at the medullary level in macaca mulatta. *Annals of Neurology, 4,* 234-244.

Strick, P.L., & Preston, J.B. (1982). Two representations of the hand in area 4 of a primate. II. Somatosensory input organization. *Journal of Neurophysiology, 48,* 150-159.

Stutsman, R. (1948). *Guide for administering the Merrill-Palmer scales of mental tests.* New York: Harcourt, Brace & World.

Tracy, D. (1980). Muscle spindle function during movement. *Trends in Neuroscience,* 257-255.

Von Hofsten, C. (1979). Development of visually directed reached: The approach phase. *Journal of Human Development Studies, 5,* 160-178.

Von Hofsten, C. (1982). Eye-hand coordination in the newborn. *Developmental Psychology, 18,* 450-461.

Von Hofsten, C., & Lindhagen, K. (1976). Observations on the development of reaching for moving objects. *Journal of Experimental Child Psychology, 28,* 158-173.

Von Hofsten, C., & Ronnquist, L. (1988). Preparation for grasping an object: A developmental study. *Journal of Experimental Physiology, 14,* 610-621.

Westling, G., & Johansson, R.S. (1984). Factors influencing the force control during precision grip. *Experimental Brain Research, 53,* 277-284.

Wise, S.D., & Evarts, E.V. (1981). The role of the cerebral cortex on movement. *Trends in Neuroscience, 4,* 297-300.

Yakoviev, P.I., & Lecours, A.R. (1967). The myelogenetic cycles of regional maturation of the brain. In A. Minkowski (Ed.), *Regional development of the brain in early life.* Oxford: Blackwell.

# 2

# Eye-Hand Coordination

*Rhoda P. Erhardt*

Skillful use of the hand under visual guidance is an important and integral part of total function. The development of eye-hand coordination skills represents a major human achievement in the ability to interact effectively with the environment and clearly reflects the capacity of the central nervous system (CNS) to receive, process, and translate visual and tactile input into efficient, well-executed motor behaviors (Paillard, 1990; Williams, 1983). An individual whose development has been delayed or compromised by damage to the CNS demonstrates fairly obvious hand dysfunction. However, visual impairments in this individual may be more subtle and difficult to identify. Comprehensive evaluations are needed to analyze the disruption of these developmental processes that interfere with the normal operation of eye-hand mechanisms.

The purpose of this chapter is to present an overview of eye-hand coordination in terms of physiological, developmental, and functional perspectives, since all three frames of reference are needed to understand the relationships between the visual and fine-motor systems. The first step is an analysis of eye-hand mechanisms, separately as well as together. A literature review then compares studies of those mechanisms relative to functional skills. A theoretical framework for evaluation and treatment begins to evolve. Specific disorders of eye-hand coordination resulting from CNS pathology and developmental delay are described, and the contents of current evaluation instruments are reviewed. The role of occupational therapy in the intervention program is delineated. Finally, a case study of a child with cerebral palsy is presented to illustrate a physiologically based, developmentally sequenced, functionally oriented operational model for intervention based on the development of eye-hand coordination in children.

## Components of Eye-Hand Mechanisms

### The Hand

The hand itself can be powerful, immobilizing an object in the palm and resisting gravity or external pull. Yet the hand can also be precise, delicately combining complicated blends of movement for independent function and for elaborate tool use. In the context of evolution, the emergence of cognitive capacities has been indicated by tool use, tool modification, and tool making (Boehme, 1988; Connolly & Dalgleish, 1989). As an instrument for perception, the hand is the source of tactile as well as proprioceptive sensations that allow identification of size, shape, texture, temperature, and weight of objects (Corbetta & Mounoud, 1990). The high density of skin receptors and wide range of possible exploratory movements provide tactile-kinesthetic spatial information that is crucial to skilled motor function.

### The Eye

Seeing is described as a function profoundly integrated with posture, manual skills, intelligence, and personality. The eye is considered a "vestibule to the brain" and "man's supreme sense" (Gesell, Ilg, & Bullis, 1949, pp. 3, 11). Information contained in the light energy entering the eye is processed against a background of information received by all other sensory systems, as well as of past experiences (Schrock, 1978). Visual perception begins with the transmission of the image from the optic structures to the visual cortex in the occipital lobe, then to the parietal lobe, where the stimulus is interpreted and differentiated, and finally directed to the frontal lobe, where the conceptual knowledge of recognition and understanding is formulated (Goble, 1984).

### The Eye-Hand Sensory System

The visual and tactile systems have very different anatomical and physiological organizations. Both sensory systems contribute similar but not identical information about objects in the environment. This information is matched and unified through cross-modal integration. To achieve harmony between visual localization and proprioceptive/tactile sensations, the direction of the eye muscle movements must agree with that of the touching limb. The hands are limited in the quantity and quality of information received, dependent on areas of the skin in actual contact with the environment. The eyes, however, can respond to stimulation from much larger and more distant parts of the environment. Thus, as the child's world broadens, the process of gathering information gradually relies more on visual than tactile input (Hatwell, 1990; Peiper, 1963).

## Functional Skills

The term *skill* refers to expertness—that is, practiced abilities showing deftness, dexterity, and confidence in functional performance. Skills are sequences of organized, goal-directed actions, which apply strategies proceeding to a future goal. Skills are influenced by the individual organism, the specific task, and the current environment. They represent a transaction between the performer and the environment, which is always changing. Therefore, flexibility and cognitive problem-solving abilities are fundamental features of skills (Connolly & Dalgleish, 1989).

It is generally accepted that basic eye-hand patterns are refined and elaborated for functional skills only through visuomotor experiences in a variety of situations (Bower, 1966; Field, 1977; Hein, 1972; McDonnell, 1979; Piaget, 1952). The performance of functional skills such as feeding, dressing, and writing requires very complex and selective patterns of muscular coordination, depending on an intact and mature CNS as well as a background of basic motor patterns that normal children acquire during the first few years of life (Bobath & Bobath, 1972).

### Hand Skills

Most studies examining hand function offer detailed descriptions of hand approach and grasp but comparatively limited information about the visual components (Connolly & Elliott, 1972; Halverson, 1931; Heinlein, 1930; Knobloch & Pasamanick, 1974; Napier, 1956; Twitchell, 1970).

Castner (1932) and Paillard (1990), however, in studies published almost 60 years apart, both recognized the importance of the eyes in hand function and described the complete act of prehension as containing three equally important phases or classes of motor operations: looking, reaching, and grasping. Castner's study of normal infants aged 20 through 52 weeks revealed well-defined patterns in all three phases. This included increases in the number and duration of regards, regards accompanied by approaches, increased accuracy of approaches, and maturity of grasps. Many children with developmental delays demonstrate only brief visual monitoring of their hands, inaccurate reaching, and immature hand grasps. Paillard used a more physiologically based approach that includes postural stabilization as a crucial factor in the early stages of acquisition. His three operations included eye-head orientation to determine correct arm and hand placement, stabilization of the trunk to ensure efficient arm transport, and controlled stability/mobility of the various arm and hand joints to achieve precise digital grasp. Children with cerebral palsy who have inadequate head and trunk control use many abnormal compensatory postures and movement patterns to reach and grasp.

Phylogenetically, both upper and lower extremities were originally used functionally for stability. Evolution of the shoulder is a prime example of the exchange of stability for mobility during the progression toward man's upright position. Changes at the proximal end of the arm became necessary as the forelimbs, no longer needed for weight bearing, were freed for prehension. The glenohumeral joint, with three axes of movement, is now the most mobile of all joints in the body. Because of its evolved shallow socket, mobility has been achieved at the sacrifice of stability. This stability is provided only by soft tissues such as muscles, tendons, and the joint capsule, rather than by bone structure (Kent, 1971).

Those soft tissues, however, provide adequate internal stability for most functional requirements of the upper extremity, which include approach, grasp, manipulation, transportation, and release of objects. Exner's (1990) descriptive model of in-hand manipulation, in particular, illustrates the importance of the stability/mobility mechanism. Various movements of objects within the hand are often accomplished while other objects are stabilized, usually by the ulnar fingers. Some children with atypical or delayed development are unable to achieve these high-level manipulative skills because they use inappropriate points of stability.

### Visual Skills

The physiology of vision is even more complex than that of prehension. Normal physiological function of the hand can be understood within the general knowledge of the body's bone structure, neuromuscular function, and tactile/proprioceptive mechanisms. However, the study of vision requires an additional understanding of the optical as well as the neural systems, and the connections between them.

The *medical* approach to visual handicaps is based on a model of pathology, which assumes that skilled visual function depends on intact anatomical structures and normal physiological processes of the eyes, visual pathways, and cerebral cortices. Dysfunction is managed by identifying the etiology of the injury or disease and remediating the pathological deviations.

The process of correlating parallel visual and prehensile data into an integrated model is facilitated by the *developmental* approach. In both areas of development, early reflexes and primitive movement patterns provide the foundation for more voluntary refined movements, and each step is dependent on the integration of the preceding steps. Developmental stages blend together imperceptibly during occasional reversions to previously outgrown levels, or during plateau periods while new skills become integrated with previous ones.

The *educational* view of visual perception is concerned with learning, adaptation, and the impact of environmental factors on function, the essence of occupational therapy. Perception is generally considered an acquired skill, defined as the recognition of sensory information. Visual perception is the capacity to interpret visual sensory input and assign meaning to what is seen. Processing of sensory information involves integration of visual data with that derived from motor and sensory receptors. Therapeutic adaptation of the environment can facilitate this integration process, directly affecting functional skills in the home, school, and community environments.

These three models are each useful for understanding the child's developing visual system. Medical data about the child describe the optical and neural physiological systems. Educational assessments and program plans state levels of function and specific objectives. Developmental reports provide sequential structures for intervention.

### Eye-Hand Skills

Psychologists and neuroscientists who have studied both manual and ocular motor systems describe the vast differences in their biomechanical properties that affect coordination of their movements. The limbs, which have considerable mass and joint freedom, move through space always affected by gravitational pull. In contrast, the eyes have a relatively negligible mass, and their movements are limited to rotation around a single point, virtually unaffected by gravitational forces (Fisk, 1990). Additionally, the eyes are centrally located near the axes of head and body, while the hands are peripheral to the proximal axis. In order to guide limb movements, the visual system uses dual processing: peripheral and central. Information from the peripheral field analyzes the trajectory of the moving limb relative to the direction of gaze. Acute central vision analyzes the relative positions of hand and target. Both processes update and correct the movement trajectory (Paillard, 1979). If the eyes and hands are to achieve accurate and coordinated function, the neural commands for their actions must be timed according to the differences in their biomechanical constraints. The temporal organization of eye and hand movement is flexible and purposeful, with both serial and parallel processing of information for the two systems. Eye movements provide retinal error signals and gaze position signals that guide hand movement. At the same time, coordinated hand movements influence eye movements by supplying information about target location. Both systems exchange information needed for adaptation to rapidly changing environmental conditions. Most importantly, eye movements normally precede hand movements to visual targets. This visual information, together with visual and tactile memory, provides both quantitative and qualitative visual information (distance, size, shape, weight, texture). The hand then moves more accurately toward the target and prepares for more precise shaping, rather than waiting for kinesthetic/proprioceptive data on contact (Fisk).

## Theoretical Framework for Eye-Hand Function

### Physiology of Eye-Hand Relationships

An important evolution of brain structures includes specialization of the parietal association cortex, a highly organized area for precise visual guidance of goal-directed arm and hand movements. Separate neural networks in different brain structures organize the functional motor operations described by Paillard (1990):

1. Eye-head orientation, which leads to identification of the object, is organized at the midbrain level around the superior colliculus.
2. Arm mobilization, which localizes the object following a triggered ballistic movement, prescribes in advance the direction and distance of travel depending on brainstem structures.

3. Grasping movement, which adjusts the terminal guided approach, is primarily under cortical control.

Each of these motor patterns receives and uses sensory information, especially visual feedback loops that have processed size and shape cues, calibrated gaze orientation, and guided arm trajectory and orientation of hand grip. After contact, tactile cues shape the grasp and assist in manipulation.

## Developmental Sequences of Eye-Hand Components

During the normal developmental process, reflexes become progressively integrated until the patterns are no longer obligatory. At this stage, they no longer interfere with voluntary control, but are still available to the individual during periods of stress. For example, the asymmetrical tonic neck reflex (ATNR), a response to head rotation, causes the face side arm to extend and the skull side arm to flex. Emerging at 1 and 2 months, ATNR helps develop visual monitoring of the hand. However, if it does not begin to diminish at 3 or 4 months and become integrated into voluntary patterns by 6 or 7 months, it may interfere with the infant's midline orientation, symmetry, bilateral hand function, and binocular vision (Gesell, Ilg, & Bullis, 1949). In a normal older child or adult, the ATNR can still be recruited for certain tasks requiring exceptional balance, accuracy, and/or strength, such as fencing, throwing, or digging.

McDonnell (1979) offered a modification of the traditional reflex-to-complex behavior model by using the term "instrumental" to describe motor behaviors that involve information processing and have a specific effect on the environment. He suggested that eye-hand activities may emerge concurrently with, rather than in transition from, the maturation of reflex functions. Instrumental behaviors may be possible from birth but occur infrequently due to the immaturity of the nervous system, which at this stage of development is particularly sensitive to physiological changes and requires optimal environmental conditions for actions such as reaching. McDonnell agreed with Piaget (1952) that reflexes provide environmental opportunities for infants to perform instrumental activities that are reinforced and repeated, and that the increase of this instrumental activity facilitates the decline of neonatal reflex functions.

Development of visual control over the hand is a major sensorimotor acquisition during the 1st half year of life. Most authors agree that visually guided reaching becomes functional by about 5 or 6 months of age, and that it peaks at 7 or 8 months. During the last half of the 1st year, the need to visually monitor the hand until reaching is completed gradually decreases. After the age of 1 year, once the hand moves efficiently toward its destination, visual attention may be transferred to another place of interest (Bruner, 1969; McDonnell, 1979; Piaget, 1952; White, 1969).

Developmental sequences of eye-hand components have been grouped in stages or phases by several different authors to describe different types of interactions. Comparisons of these varied methods of organizing information provide an important awareness of how the same subject is treated from different frames of reference. As different terminology is used to describe similar behaviors, the total picture is enhanced and better understood. In Figure 1, Holt (1977) emphasizes the importance of tactile sensations for early visual exploration and awareness.

Phases of reaching-prehensile behavior requiring the collaboration of visual and motor mechanisms were described by McGraw (1969), based on observations of children aged birth to 4 years. These phases included newborn-passive, object-vision, visuomotor, manipulative-deliberate, visual release, and mature (see Table 1).

White, Castle, and Held (1964) studied infants under 6 months of age and categorized eight stages of the normative developmental sequence of visually directed reaching (see Table 2). Distinct sensorimotor systems, including visuomotor of eye-arm and eye-hand and tactual-motor of the hands, contribute to the development of prehension and gradually become coordinated and integrated. During stages one and two, the growing capacity of the infant to localize and follow with eyes and head is an important prerequisite for visually directed prehension, since tactual hand activities are operating in isolation from eye and head movements. During stages three and four, the eyes can focus on objects within the infant's reach, and time is spent watching his or her own hands. Thus movements of the eyes, head, and arms begin to coordinate in swiping. Grasping is tactually directed only. Alternating glances, however, allow the visuomotor systems of eye-object and eye-hand to be compared. During stages five and six, spontaneous mutual fingering in midline is combined with visual monitoring, linking vision and touch by a double feedback system. Each hand is simultaneously touching and being touched, and the eyes are seeing what the hands are feeling. During stages seven and eight, integration of eye-hand, eye-object, and tactual-motor is accomplished, resulting in adultlike visually directed reaching and grasping.

Because the development of highly refined finemotor behaviors is a process that requires the organism to simultaneously see and feel hand movements during interaction and exploration of the environment, vision serves an important regulating function. Held and Bauer (1967) reported that a rhesus monkey not allowed to see its hands during its 1st month

**Figure 1. Developmental Sequences of Visual and Tactile Linkages (Source: Holt, 1977)**

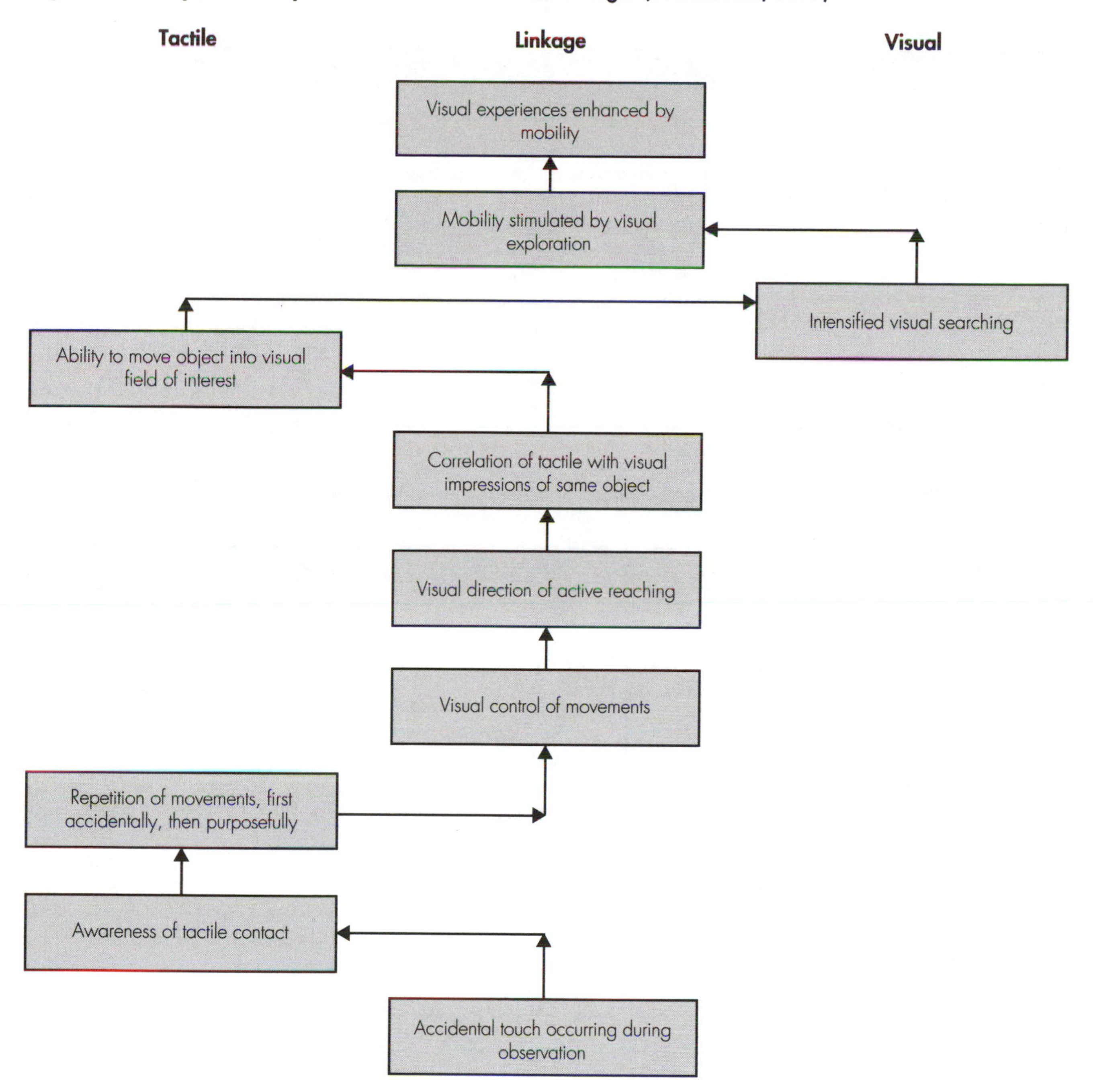


did not reach accurately for an object when its hand was first exposed, but instead gazed at the hand as it entered the visual field. Accurate visual guidance of reaching and grasping was then eventually achieved, suggesting that integration of nonvisual limb control and visuomotor head control first required viewing of the moving hand.

A study of the early development of eye-hand coordination in humans from the perspective of exercise science emphasized the major role of the visual mechanism (Williams, 1973). Stages beginning at birth and continuing to 40 weeks and beyond were described as (1) static visual exploration, (2) active and repeated visual exploration of objects, (3) visual regulation of grasping/manipulation, and (4) refined control of eye-hand behaviors (see Table 3).

As tactile/proprioceptive sensory input stimulates reflexive grasp patterns, the eyes begin to explore the environment, including the hands. Vision helps to integrate both reflexive and voluntary movement systems, bringing eye-hand coordination to higher skill levels. Table 4 represents a compilation of several authors' work, correlating the stages of eye function, hand function, and their interrelationship from 1 to 6 months of age.

**Table 1. Reaching-Prehensile Behavior (Source: McGraw, 1969)**

| Phase | Description |
|---|---|
| A. Newborn-passive | No overt motor response to object within visual field; reflexive grasp on object placed in hand; no connection between visual and neuromuscular movements. |
| B. Object-vision | Visual fixation within near range associated with increase or decrease of diffuse motor activity. |
| C. Visuomotor | Approach movements of upper extremeties and digits evoked by visual gaze; connections between visual and neuromuscular interaction without purposeful activity. |
| D. Manipulative-deliberative | Sustained visual attention during voluntary movement toward object. |
| E. Visual release | Sustained visual attention not needed when reaching for object in immediate visual field. |
| F. Mature | Both visual and neuromuscular performances reduced to minimum essentials. |

**Table 2. The Normative Sequence of Visually Guided Reaching (Source: White, Castle, & Held, 1964)**

| Stage | Age in Months | Description |
|---|---|---|
| One | 1 to 1.5 | Brief or no visual response; no reach. |
| Two | 1.5 to 2 | Brief glance, sustained on side of ATNR; increased activity if inactive, decreased activity if active. |
| Three | 2 to 2.5 | Visual fixation and swiping; fisted hand. |
| Four | 2.5 to 3 | Visual fixation, accurate swiping, and alternating glances between hand and object. |
| Five | 3 to 3.5 | Bilateral hand activity, raised or clasped in response to midline presentation, or one hand raised and other clutching shirt to side presentation, and alternating glances. |
| Six | 3.5 to 4 | Bilateral responses predominating, sustained hand regard and visual monitoring. |
| Seven | 4 to 4.5 | Bilateral and unilateral responses, both hands raised or one hand encountering object, opening in anticipation, crudely grasping, alternating glances. |
| Eight | 4.5 to 5 | Rapid lifting of one hand from out of the visual field directly toward object, opening in anticipation, alternating glances, successful grasp. |

**Table 3. Early Development of Eye–Hand Coordination (Source: Williams, 1973)**

| Age in Weeks | Stage | Description |
|---|---|---|
| 0 to 16 | Static visual exploration | Intent fixation on hands, alternating with objects; eyes immobilize with objects; eyes immobilize while arms and legs activate. |
| 17 to 28 | Active visual exploration | Repeated ocular grasping and manipulation of objects. |
| 28 to 40 | Visual regulation of grasping/manipulation | Adjustment/correction of reaching during visual fixation. |
| 40+ | Refined control of eye-hand behaviors | Extension of performance to wide variety of tasks. |

**Table 4. The Development of Eye–Hand Coordination (Visual/Prehension Schemas) (Sources: Erhardt, 1982a&b; Flavell, 1963; McGraw, 1969)**

| Age | Eyes | Hands | Interrelationship |
|---|---|---|---|
| Natal | Random eye movements and passive regard of surroundings | Random and reflexive hand movements | No visual attention to hands |
| 2 months | Attracted by movement in periphery (surroundings); monocular or binocular fixation | Arms activate in response to stimulus; fingers open and close reflexively | Regards hand in ATNR position only with one active eye; releases gaze to surroundings |
| 3 months | Localizes noisy, illuminated, or moving targets | Swipes and contacts objects at side, not midline | Briefly regards own moving hand spontaneously; watches own hand reach and contact object |
| | Tracks with difficulty through 180°; converges and diverges on targets moving toward and away | Sustains grasp on objects placed in hand | Visually searches for object at point of disappearance but cannot combine reach and grasp |
| | Begins vertical tracking downward, losing target | Releases object involuntarily with awareness after sustained grasp | No retrieval of dropped or lost object |
| | Shifts gaze between two targets in same focal length | | Alternates glances from hand to object |
| 4 months | Binocular fixation | Reaches with both hands, contacts, and pulls object back against body or into mouth | Brings object to mouth without visual monitoring, then relocates after removing |
| | Prolonged, selective fixation on target in midline | Midline fingering | Maintains fixation on own hand or object in hand |
| | Visually pursues lost target outside visual field | Does not reach for lost object | Retrieval not possible when hand and object are not in same visual field |
| 5 months | Fixates on tiny target (pellet) | No attempt to grasp pellet | Visually attends to pellet but doesn't approach |
| | Visually pursues lost target outside visual field | Hands pursue lost object | Tries to retrieve lost object outside visual field |
| | Jerky vertical and diagonal tracking | Reaches with hands to corral, grasp, and maintain grasp while shaking object, shoulder motion only | No visual attention to hand shaking (manipulating) |
| | Shifts gaze between targets in near and middle space | | Shifts gaze (releases fixation) while grasping, manipulating, and mouthing target |
| 6 months | Localizes large and small targets | Reaches and grasps with one hand and shakes object (manipulates it) | Alternately looks at, mouths, shakes object, retaining grasp |
| | Fixates on pellet; fixates intensely | Contacts and rakes pellet | Fixates on pellet while reaching, contacting, and raking |
| | Visually pursues lost target outside visual field unless distracted by another within visual field | Adjusts body position to reach object outside visual field | Searches for lost object with eyes, hand, and body movements |
| | Shifts gaze in same focal length and within different focal lengths | Alternately grasps, mouths, looks, drops, grasps, transfers, and shakes object; shoulder and elbow motion | Shifts gaze during adaptive interaction of eyes, mouth, and hands |

One component of a theoretical framework for eye-hand function—physiology—has served as a structural base for a second component—developmental sequences. The third component of this theoretical framework is its functional importance.

## Functional Relevance of Eye-Hand Skills

Self-help activities (feeding, dressing, and grooming) represent major functions of the child in the home. Eye-hand components of feeding begin when the eyes locate the bottle, which leads to the hands bringing it to the mouth at approximately 6 months (age ranges vary because of individual differences). Finger feeding (10 months) and utensil manipulation (spoon and cup: 18 months; fork: 28 months; knife: 6 years) require the eyes to direct the hands for precise movements. Eventually, experience with a variety of tools in meal preparation leads to increasing independence. Dressing skills begin with undressing (2-3 years) and dressing (3 years). Ability to dress the lower body (easier visual and manual access) usually emerges before ability to dress the upper body. Fasteners (buttons, zippers, snaps, and buckles), which require sustained visual monitoring and a high level of manual dexterity, are not usually attempted with success until 3 years; whereas ability to tie shoelaces emerges at 4 or 5 years of age. Most grooming abilities emerge between 3 and 4 years and include washing hands and face (mirror awareness), bathing (visually scanning body as well as monitoring hands), toothbrushing, hair care (mirror and use of brush/comb in hand), toileting (cleansing self), and, finally, nail care, probably not until 5 or 6 years (Coley, 1978). During adolescence, interest in appearance surges, as cosmetics, hair styles, and clothes preoccupy both boys and girls. These grooming skills require an exceptional degree of visual inspection and eye-hand coordination.

Play, which is a primary occupation of childhood, is another important function in the home environment. Play is characterized by fun and spontaneity and involves exploration, experimentation, imitation, and repetition of experience. It integrates the child's internal and external world and follows a sequential, developmental progression (Florey, 1981). Piaget (1962) described stages of cognitive play approximating stages of intellectual development, which he defined as adaptation to new situations: functional or practice games during the sensorimotor period (0-2 years); construction and symbolic games during the preoperational period (2-7 years); collective games with rules during the concrete operations period (7-11 years); and formal operations period (12+ years). His schemas of object permanence, means-end abilities, vocal imitation, gestural imitation, operational causality, spatial relationships, and scheme actions are the sensorimotor domains of the Uzgiris and Hunt Scales of Psychological Development (Uzgiris & Hunt, 1975). Systematic methods of observing play behavior include developmental sequences of physical as well as cognitive and social aspects (Bledsoe & Shepherd, 1982; Hulme & Lunzer, 1966; Knox, 1974). The Parten Social Play Hierarchy examines the degree of participation in play with descriptors such as unoccupied behavior, solitary play, onlooker behavior, parallel play, associative play, and cooperative play (Parten, 1932). The Preschool Play Scale (Bledsoe & Shepherd; Knox) contains dimensions of material management, space management, imitation, and participation, with categories within each dimension, such as exploration, interest, and imitation described in yearly increments from birth to 6 years.

These play scales can be valuable tools for pediatric therapists working with children who have disabilities. Children develop problem-solving skills for future independent life while attending to environmental consequences of their own actions during play. Through its imaginary or quasi-reality nature, play allows children to create new situations without fear of repercussions. Therapists can stimulate increased play by providing safeness and limited consequences for risk taking, as well as their own playful attitude (Vandenberg & Kielhofner, 1982). Eyes and hands are both involved in activities such as rotating and examining three-dimensional aspects of objects, inserting objects into small openings, discriminating and matching similar objects, constructing complex structures, and using tools such as hammers, brushes, and pencils (Linder, 1990).

The scribbles and drawings of children, described as a type of "motor" babbling, are influenced by both anatomical and perceptual factors. Although the straight line is visually simplest, it requires complex muscular control by the multiple-jointed arm. The first scribbles are actually angular zig-zag lines, which are related to the lever constructions of the arm joints. They are gradually smoothed to become circular strokes. Circles, which appear often in children's scribblings as well as in early cave drawings, represent a primitive visual symbol: the eyes of animals and people (Arnheim, 1974; Cratty, 1986). During early use of writing implements, only the pencil tip touches the page, as the arm and hand move unanchored through space. Trunk and shoulder are providing internal stability for arm mobility. Later, the elbow and ulnar side of the hand rest on the writing surface (external stabilization) as forearm mobility depends on elbow stability. Next, the wrist stabilizes on the surface and the hand moves. Finally, with maturity, the hand itself stabilizes, allowing finger movement dissociated from the metacarpophalangeal (MCP) joints. This developmental progression of the stabilizing point of control from proximal to distal joints has been correlated with that of immature (externally stabilized) to mature (internally stabilized) pencil grips (Erhardt, 1982a&b). The role of vision is to reinforce accidental efforts, slow down movements during attempts to visually guide them, and to keep the child's attention on the task by sustained visual tracking of the hand (Cratty, 1986). Visual feedback is used more efficiently in older children. For example, Schneck & Henderson (1990) reported that older children slowed down when coloring near the edge of a circle, while younger children achieved accuracy by changing to a less mature but more stable grip.

Once the child enters school, functional eye-hand skills become even more important for writing, sports, and prevocational activities such as keyboard use. Scribbling, coloring, and drawing have provided the visuomotor experiences that are necessary for complex writing tasks. These tasks are learned by tracing (eyes direct hand to follow the visual representation), imitating (eyes watch and remember another's action in order to repeat the same action and production), and copying (eyes alternate glances between visual representation and own production in process). Games and sports require even higher levels of eye-hand coordination for efficient throwing, catching, and hitting of a variety of balls. Keyboards and computers that make specific demands on the visuomotor system at school and home also offer some occupational choices for the child growing into adulthood. Functional skill requirements involving tools contain components of abilities that are innate but also must be practiced to reach a safe level of competency, e.g., operating a power saw, driving a car, or performing surgery (Starkes, 1990).

## Disorders of Eye-Hand Coordination

Pathology and/or developmental delay of either the central or peripheral neuromuscular systems can result in eye-hand dysfunction. Obvious orthopedic conditions affecting the hands, such as juvenile rheumatoid arthritis, arthrogryposis, amputations, or congenital anomalies, will alter the course of development and/or functional dexterity, even if the visual system is intact. Conversely, the effects of visual impairment on hand development have been well-documented. Fraiberg (1971) described children blind from birth as having "blind" hands, hands that do not explore objects but only serve to bring them to the mouth. In the absence of vision, the sensorimotor sequence that leads to adaptive hand behavior may not evolve. Children with low vision are less likely to see objects that trigger spontaneous exploration; their experience is thus limited (Harrell & Akenson, 1987).

Most developmental disabilities are a result of CNS damage that affects both eye and hand function. The original pathology is usually a static event, not progressive, but the results of this pathology have significant implications for the course of development.

For example, children with cerebral palsy are not only motorically delayed but are hampered by poor sensory feedback from muscle and joint proprioceptors, influencing motor output. Some of these children appear to use vision to compensate for inadequate and inaccurate proprioception, but others seem to ignore visual input, possibly because processing data from multiple channels is too difficult. Data may also be distorted by ocular problems such as refractive errors, strabismus, or nystagmus.

Visual, perceptual, and upper-extremity problems are also seen with spina bifida. Fine-motor delays in these children can be caused by decreased strength, reduced reaction time, inefficient motor planning, and poor manipulation skills (especially bilateral), with serious impact on functional abilities. Upper extremity control is needed for feeding, dressing, and handwriting, as well as mobility (crawling, transitions, and crutch use). Perceptual dysfunction has been associated with hydrocephalus. Specific deficits are found in processing of tactile input, visual attention to task, spatial judgments, and eye-hand coordination. Visual impairments may be in the eye itself or in the processing of visual information. Refractive errors, strabismus, and nystagmus are found in children with myelomeningocele as well as those with cerebral palsy, interfering with the acquisition of many eye-hand skills. The learning process is affected by errors in visual judgments during reaching, distortion of spatial relationships, and compensatory head posturings (Williamson, 1987).

Children with learning disabilities comprise another large group who demonstrate problems with the integration of motor and visual information. Many have visuomotor problems that can be related to visual perceptual problems and/or motor planning difficulties. These problems can interfere with skilled upper-limb movements, including tool use and activities of daily living.

## Assessment

The assessment and subsequent treatment of fine-motor dysfunction are major areas of expertise for occupational therapists. But they cannot be fully addressed without consideration of the visual components of prehension.

Many professionals working in the field of developmental disabilities have created checklists, charts, models, and scales to help clarify the developmental progression of skilled hand use. Many of these have included visual components. For example, *A Model of Sensorimotor Systems in Development of Reach and Prehension During the First 12 Months of Life* (Ammon & Etzel, 1977) depicts the temporal relationship of several systems that culminate in mature hand skills, including visual responses of attention, pursuit, hand regard, and visually directed reaching. *The Development of Reach and Grasp* (Cliff, 1979) evaluation chart examines regard as well as posture, approach, grasp, manipulation, and release. Test items are grouped for scoring purposes as specific eye-hand coordination skills within the fine-motor scale of the *Peabody Developmental Motor Scales* (Folio & Fewell, 1983). Eye-hand behaviors are also included in a developmental grid accompanying *The Developmental Therapist* (Banus, Norton, Sukiennicki, 1979). The grid is designed in a continuum to clarify the differences between sensory/perceptual development and behavioral responses by considering input, integration, and output separately. *The Erhardt Developmental Prehension Assessment* (Erhardt, 1982b) includes progression of hand/vision/sucking schemas in the manipulation cluster, while the *Erhardt Developmental Vision Assessment* (Erhardt, 1989) shows sequences of eye-hand interaction in the localization, fixation, and gaze shift clusters. A toy selection guide for developmentally delayed children analyzes arm and hand functions necessary for each age-appropriate toy and contains three charts: hand development, use of arms together, and hand-eye coordination (Frantzen, 1957). The *Preschool Play Scale* (Knox, 1974) measures visual as well as manual exploration within dimensions of space and material management.

Psychological and cognitive evaluations often address the connections between eyes and hands. *The Uzgiris and Hunt Scales of Infant Psychological Development* (Uzgiris & Hunt, 1975) list critical behaviors such as securing hidden objects within the context of visual displacement and pursuit, changing activity level linked with visual awareness, visually inspecting objects in hand, shaping the hand during visually directed reaching, imitating visible gestures, watching the hand and repeating arm movements in response to an interesting spectacle, and using hand procedures to create causal behavior.

Certain evaluations of sensory processing, such as the *Sensory Integration and Praxis Test* (Ayres, 1989) and the*Test of Sensory Function in Infants* (DeGangi & Greenspan, 1989), include specific sections measuring tactile perception and the integration of vision. *The Sensorimotor Performance Analysis* (Richter & Montgomery, 1989) scores visual tracking, visual avoidance, visual processing, and eye-hand coordination during gross- and fine-motor tasks.

A unified comprehensive measure of eye-hand coordination, which is developmentally based and functionally oriented, would be useful for therapists who assess individuals of all ages with developmental disabilities. In order to assist these individuals to achieve their highest level of independence, therapists must obtain specific information about those integrated skills. Programs can then be directed toward a reasonable balance of developmentally targeted interventions and functionally appropriate adaptations. Younger children or children with milder problems will respond to developmental treatment more rapidly and with more gains, while older children or more severely involved children will need more environmental adaptations in order to perform eye-hand tasks.

## Intervention

### Role of Occupational Therapy

The fundamental concern of occupational therapy is the capacity of the individual to perform those tasks and roles essential to productive living. The child's primary occupational role is play, a highly motivating adaptive process, engaged in for its own sake. The therapist's unique therapeutic use of play requires the reconceptualization of play as a purposeful activity. An important goal of occupational therapy in treating a child with cerebral palsy is to promote normal patterns of movement and prevent abnormal postural reactions while the child participates in functional, purposeful activity, including play. The neurodevelopmental treatment (NDT) method developed by Berta Bobath and Dr. Karel Bobath is commonly used by pediatric therapists to elicit those active, normal responses (Boehme, 1988). Play is characterized by three types of behaviors: (1) exploratory behaviors (motivated by intrinsic curiosity); (2) competency behaviors (mastery gained by practicing skills that have demonstrated cause and effect); and (3) achievement behaviors (meeting standards of excellence externally motivated). Activity adaptation, which is inherent in the effective use of play, can be accomplished by (1) adapting the size or shape of materials and equipment, (2) modifying the procedures, (3) adjusting the position of child or materials, and (4) controlling the nature of interpersonal interaction (Anderson, Hinojosa, & Strauch, 1987).

Another important occupational role of childhood is that of student or learner, not only of academics in the school environment, but also of self-help skills to increase independence in the home and community (Hopkins & Smith, 1978). NDT strives to help the child produce automatic movement patterns without placing conscious attention on the process. The child absorbed in play or learning a new task is not focused on the activity's motor demands.

A therapeutic environment must be created for the individual in which the required behaviors and skills can be practiced and learned. From a physiological perspective, motor output is dependent on sensory input and can be influenced by intervention, which utilizes sensory stimuli (visual, auditory, tactile, proprioceptive, vestibular, olfactory, and gustatory) to activate normalized motor responses of the eyes and the hands in developmental sequences. Occupational therapists are also trained to adapt environmental conditions (equipment, positioning, etc.) to the individual who, growing older, needs a shift of emphasis from the developmental approach to an increasingly functional focus. Therefore, the occupational therapist should draw upon a physiologically based and developmentally sequenced body of information when designing assessment and intervention strategies for children with functional eye-hand coordination problems.

### An Operational Model for Eye-Hand Coordination Intervention

A proposed model for intervention presents a systems approach to the assessment and treatment of eye-hand coordination deficits in children. Both developmental and functional approaches are interwoven in the model for identifying needs and planning intervention strategies. The pediatric occupational therapist first considers the child's intrinsic motivation in the context of cognitive development and play schemas, which ultimately results in purposeful, goal-directed, eye-hand coordination behaviors. Exploratory stages including visual exploration, tactile exploration, and eye-hand interaction, fit within the physiological frame of reference as specific developmental sequences are used to implement the process. The therapist links these developmentally targeted interventions with functionally appropriate adaptations for practical consideration of theoretical concepts, including proximal/distal development and stability/mobility factors (see Table 5).

## CHAD: A CASE STUDY

The following case study illustrates an application of the proposed model for a young child with cerebral palsy.

### Medical History and Diagnoses

Chad is a 5-year-old male with cerebral palsy, developmental delay, and a seizure disorder, which is controlled with medication. Prenatal complications included maternal hypertension, severe preeclampsia, and fetal distress, resulting in a cesarean section delivery at full-term. Birth weight was 6 lbs., 6 oz., with Apgar scores of 2 and 9. Neonatal complications included hypoglycemia and respiratory distress syndrome. Postnatal complications included esophageal reflux and infantile spasms. Visual impairments include significant hyperopia (farsightedness), astigmatism, nystagmus, and a right exotropia (outward eye turn).

### Gross- and Fine-Motor Development

Chad can roll and belly crawl but cannot sit, stand, or walk alone. Low muscle tone proximally (neck and trunk) contributes to inconsistent head control, particularly in downward gaze, interfering with eye-hand

**Table 5. An Operational Model for Eye–Hand Coordination (Sources: Coley, 1978; Erhardt, 1982a&b; Erhardt, 1990; Frantzen, 1957; Uzgiris & Hunt; 1975)**

| System Components | Developmentally Targeted Interventions | Functionally Appropriate Adaptations |
|---|---|---|
| Cognitive motivation ▼ | Piaget's schemas; developmental play sequences | Adapted age-appropriate toys; switches; adapted positioning (supine, prone, sitting, sidelying, standing) |
| Visual exploration ▼ | Environmental awareness (near, middle, and far space; central and peripheral fields); linkage with other stimuli (sound, movement) | Adapted materials, lighting, and positioning (trunk support for eye–head mobility, head support for eye mobility) |
| Tactile exploration ▼ | Developmental sequences of approach, grasp, manipulation, release; touching objects within reach, experiencing size, shape, texture, temperature | Adapted materials, equipment, and positioning (trunk support for arm mobility, table surface for hand/finger control) |
| Eye–hand interaction ▼ | Visually directed exploration, repetition, adaptation | Adapted seating and positioning to ensure both hand and object in visual field; appropriate proximal stability for distal mobility |
| Purposeful activity | Goal-directed, problem-solving manipulation, self-help activities, tool uses, prewriting developmental sequences | Task analysis, adapted utensils, graded work surfaces, low-tech aids, high-tech electronic equipment |

coordination. Muscle tone increases in the distal extremities during effort or excitement. According to informal neurodevelopmental treatment (NDT) evaluation, Chad has problems with graded movement and midline control. He appears to rely on his left side for stability and his right side for mobility, which limits left arm movement and use of both hands together.

Reaching and grasping a variety of objects is a functional skill, with the right hand performing better than the left. Manipulation skills include pushing, pulling, shaking, and throwing with limited tactile exploration, bilateral play, transferring, and eye-hand coordination. Since release into containers or on a surface is difficult, Chad prefers to drop, throw, or release into another's hand. Prewriting skills need adapted materials and maximum assistance. The Erhardt Developmental Prehension Assessment (EDPA), which measures hand function from the fetal and natal periods to 15 months, was used to determine Chad's fine-motor development. The 15-month level is considered to be the maturity of prehension (and thus an appropriate norm for older children), since essential pattern components are functional, and further refinement, increased skill, and tool use result from learned experiences. According to the EDPA, Chad's fine-motor development ranges from 1 to 10 months in the left hand, and 2 to 15 months in the right hand.

### Visuomotor Development

The Erhardt Developmental Vision Assessment (EDVA), which measures the motor components of vision from the fetal and natal periods to 6 months, was used to determine Chad's visuomotor development. The 6-month level is considered to be a significant stage of maturity (and thus an appropriate norm for older children), since primitive reflexes are integrated and essential eye movement components are nearly as functional as in the adult. According to the EDVA, visuomotor development ranges from birth to 6 months. Although Chad is considered legally blind (visual acuity less than 20/200 with correction), he has functional residual vision. His corrective lenses are designed to help him achieve focus at a 16-22" working distance. External head support is essential for eye-hand activities requiring visual fixation, pursuit, or gaze shift, since Chad does not have the internal head control needed as a stable base and to dissociate eyes from head. Inclined table or desk surfaces are important for downward gaze with eyes and face parallel to the visual material.

### Positioning and Adaptive Equipment

Chad's travel and power chairs both incline his trunk and provide support for his head in a vertical position. A vest is used to stabilize his upper trunk. Neoprene hand splints assist thumb opposition. A dowel on the tray stabilizes his left hand during right hand use, and facilitates midline orientation and symmetry.

### Cognitive Development

Chad is currently mainstreamed into a regular Montessori preschool classroom. According to the parent and the school reports, he demonstrates cognitive skills ranging from Piaget's sensorimotor period (0-2 years) through the preoperational period (2-7 years), with some behaviors leading to the concrete operations period (7-11 years). For example, he uses an adapted switch to operate a remote-controlled toy car (sensorimotor period: causality, means-end). He plays independently with velcro blocks and tray (preoperational period: construction). He makes choices with an eye-gaze board to play card games (concrete operations period: collective games with rules). Chad also demonstrates a variety of physical, social, and cognitive aspects of play behaviors. His scores on the Preschool Play Scale (Knox, 1974) range from 1 to 6 years of age. For example, he enjoys driving his power chair around his neighborhood to find and touch objects such as trees and fire hydrants (space management dimension, including exploration and territory: interest in new environmental experiences and physical area used during play). He also enjoys playing with dolls (imitation dimension including dramatization and imagination: pretending, assuming roles, and introducing novelty into a situation).

### Occupational Therapy Home Program

Table 6 (see page 28) illustrates the practical application of the model for Chad's occupational therapy home program, implemented by both parents (see Figure 2).

The role of the occupational therapist includes detailed assessment of both prehension and vision, participation in the planning of both home and school programs, and coordination of those programs through periodic consultations with all team members, including parents.

Chad's activities, which include cognitive, visual, tactile, eye-hand, and purposeful components, are accomplished in a variety of postures, such as prone, sidelying, sitting, and standing. Positioning, equipment, and materials are adapted to meet his specific needs, including enhancement of low vision, head and trunk stability, arm mobility, and eye-hand coordination. The social aspects of parent involvement, play, and self-help activities are addressed, blended with developmentally targeted interventions and functionally appropriate adaptations.

## Summary

Components of eye-hand mechanisms have been described through a review of studies concerning hand skills, visual skills, and eye-hand skills within physiological, developmental, and functional frames of reference. Disorders of eye-hand function due to CNS pathology and/or developmental delay can be evaluated by a variety of test instruments containing sections addressing eye-hand coordination. However, a unified, comprehensive assessment of eye-hand coordination, developmentally based and functionally oriented, would be useful for individuals of all ages who have developmental disabilities. The role of the occupational therapist has been described as facilitator of play and learning, important occupations of childhood leading to productive living. A case study of a 5-year-old child with cerebral palsy has illustrated a proposed model for intervention that suggests practical management methods, including both developmentally targeted interventions and functionally appropriate adaptations.

**Figure 2. Chad**

## Acknowledgment

Special thanks are extended to Chad, Valerie, and James Sand for their gracious cooperation and assistance.

## References

Ammon, J.E., & Etzel, M.E. (1977). Sensorimotor organization in reach and prehension: A developmental model. *Physical Therapy, 57*(1), 7–14.

Anderson, J., Hinojosa, J., & Strauch, C. (1987). Integrating play in neurodevelopmental treatment. *American Journal of Occupational Therapy, 41*(7), 421–426.

Arnheim, R. (1974). *Art and visual perception.* Berkley, CA: University of California Press.

Ayres, A.J. (1989). *Sensory integration and praxis tests.* Los Angeles: Western Psychological Services.

Banus, B.S., Kent, C.A., Norton, Y., & Sukiennicki, D.R. (1979). *The developmental therapist.* Thorofare, NJ: Slack.

Batshaw, M.L., & Perret, Y.M. (1981). *Children with handicaps: A medical primer.* Baltimore, MD: Paul H. Brookes.

Bledsoe, N.P., & Shepherd, J.T. (1982). A study of reliability and validity of a preschool play scale. *American Journal of Occupational Therapy, 36*(12), 783–788.

Bobath, K., & Bobath, B. (1972). Cerebral palsy. In P.H. Pearson & C.E. Williams (Eds.), *Physical therapy services in developmental disabilities* (pp. 31–185). Springfield, IL: Charles C. Thomas.

Boehme, R. (1988). *Improving upper body control.* Tucson, AZ: Therapy Skill Builders.

Bower, T.G.F. (1966). The visual world of infants. *Scientific American, 215,* 80–92.

Bruner, J.S. (1969). Eye, hand, and mind. In D. Elkind & J. Flavell (Eds.), *Studies in cognitive development* (pp. 223–235). New York: Oxford University Press.

Castner, B.M. (1932). The development of fine prehension in infancy. *Genetic Psychology Monographs, 12*(2), 105–191.

Cliff, S. (1979). *The development of reach and grasp.* El Paso, TX: Guynes Printing.

Coley, I.L. (1978). *Pediatric assessment of self-care activities.* St. Louis, MO: Mosby.

Connolly, K., & Dagleish, M. (1989). The emergency of tool-using skill in infancy. *Developmental Psychology, 25*(6), 894–912.

Connolly, K., & Elliott, J. (1972). The evolution and ontogeny of hand function, In N. Blurton-Jones (Ed.), *Ethological studies of child behavior* (pp. 329–383). Cambridge: University Press.

Corbetta, D., & Mounoud, P. (1990). Early development of grasping and manipulation. In C. Bard, M. Fleury, & L. Hay (Eds.), *Development of eye–hand coordination across the life span* (pp. 188–216). Columbia, SC: University of South Carolina Press.

Cratty, B.J. (1986). *Perceptual and motor development in infants and children.* Englewood Cliffs, NJ: Prentice-Hall.

DeGangi, G., & Greenspan, S.I. (1989) *Test of sensory functions in infants.* Los Angeles: Western Psychological Services.

Erhardt, R.P. (1982a). *Developmental hand dysfunction: Theory, assessment, treatment.* Tucson, AZ: Therapy Skill Builders.

Erhardt, R.P. (1982b). *The Erhardt developmental prehension assessment (EDPA).* Tucson, AZ: Therapy Skill Builders.

Erhardt, R.P. (1989). *The Erhardt developmental vision assessment (EDVA).* Tucson, AZ: Therapy Skill Builders.

Erhardt, R.P. (1990). *Developmental visual dysfunction: Models for assessment and management.* Tucson, AZ: Therapy Skill Builders.

Exner, C.E. (1990). The zone of proximal development in in-hand manipulation skills of nondysfunctional 3- and 4-year-old children. *American Journal of Occupational Therapy, 44*(10), 884–892.

Field, J. (1977). Coordination of vision and prehension in young infants. *Child Development, 48,* 97–103.

Fisk, J.D. (1990). Sensory and motor integration in the control of reaching. In C. Bard, M. Fleury, & L. Hay (Eds.), *Development of eye–hand coordination across the life span* (pp. 75–98). Columbia, SC: University of South Carolina Press.

Flavell, J. (1963). *The developmental psychology of Jean Piaget.* Princeton, NJ: Van Nostrand.

Florey, L.L. (1981). Studies of play: Implications for growth, development, and for clinical practice. *American Journal of Occupational Therapy, 35,* 519–528.

Folio, M.R., & Fewell, R.R. (1983). *Peabody developmental motor scales.* Allen, TX: DLM Teaching Resources.

Fraiberg, S. (1971). Intervention in infancy: A program for blind infants. *Journal of the American Academy of Child Psychiatry, 10,* 381–405.

Frantzen, J. (1957). *Toys: The tools of children.* Chicago: National Society for Crippled Children and Adults.

Gesell, A., Ilg, F.L., & Bullis, G.E. (1949). *Vision: It's* [sic] *development in infant and child.* New York: Paul B. Hoeber.

Goble, J.L. (1984). *Visual disorders in the handicapped child.* New York: Marcel Dekker.

Halverson, H.M. (1931). An experimental study of prehension in infants by means of systematic cinema records. *Genetic Psychologic Monographs, 10,* 107–286.

Harrell, L., & Akenson, N. (1987). *Preschool vision stimulation: It's more than a flashlight!* New York: American Foundation for the Blind.

Hatwell, Y. (1990). Spatial perception by eyes and hand: Comparison and intermodal integration. In C. Bard, M. Fleury, & L. Hay (Eds.), *Development of eye-hand coordination across the life span* (pp. 99–132). Columbia, SC: University of South Carolina Press.

Hein, A. (1972). Acquiring components of visually guided behavior. In A.D. Pick (Ed.), *Minnesota symposia on child development* (pp. 53–68). Minneapolis: University of Minnesota Press.

Heinlein, J.H. (1930). Preferential manipulation in children. *Comparative Psychology Monographs, 7,* 1–121.

Held, R., & Bauer, J. (1967). Visually guided reaching in infant monkeys after restricted rearing. *Science, 155,* 718–720.

Holt, K.S. (1977). *Developmental pediatrics.* London: Butterworth.

Hopkins, H.L., & Smith, H.D. (1978). *Willard & Spackman's occupational therapy, 7th edition.* Philadelphia: Lippincott.

Hulme, I., & Lunzer, E. (1966). Play, language, and reasoning in subnormal children. *Journal of Child Psychology, 7,* 107-123.

Kent, B.E. (1971). Functional anatomy of the shoulder complex. *Physical Therapy, 51*(8), 867-889.

Knobloch, H., & Pasamanick, B. (Eds.). (1974). *Gesell & Amatruda's developmental diagnosis.* Hagerstown, MD: Harper & Row.

Knox, S. (1974). A play scale. In M. Reilly (Ed.), *Play as exploratory learning* (pp. 247-266). Beverly Hills, CA: Sage Publications.

Linder, T.W. (1990). *Transdisciplinary play-based assessment.* Baltimore: Paul H. Brookes.

McDonnell, P.M. (1979). Patterns of eye-hand coordination in the first year of life. *Canadian Journal of Psychology, 33*(4), 253-267.

McGraw, M.B. (1969). *The neuromuscular maturation of the human infant.* New York: Hafner Publishing.

Napier, J.R. (1956). The prehensile movements of the human hand. *Journal of Bone and Joint Surgery, 38B,* 902-913.

Paillard, J. (1979). Distinctive contribution of peripheral and central vision to visually guided reaching. In D. Ingle, M.A. Goodale, & R.J.W. Mansfield (Eds.), *Analysis of visual behavior* (pp. 367-385). Cambridge, MA: MIT Press.

Paillard, J. (1990). Basic neurophysiological structures of eye-hand coordination. In C. Bard, M. Fleury, & L. Hay (Eds.), *Development of eye-hand coordination* (pp. 26-74). Columbia, SC: University of South Carolina Press.

Parten, M. (1932). Social participation among preschool children. *Journal of Abnormal Social Psychology, 27,* 243-269.

Peiper, A. (1963). *Cerebral function in infancy and childhood.* New York: Consultants Bureau.

Piaget, J. (1952). *The origins of intelligence in children.* New York: International University Press.

Piaget, J. (1962). *Play, dreams, and imitation in childhood.* New York: W.W. Norton.

Richter, E.W., & Montgomery, P.C. (1989). *The sensorimotor performance analysis.* Hugo, MN: PDP Press.

Rosenbloom, L., & Horton, M.E. (1971). The maturation of fine prehension. *Developmental Medicine & Child Neurology, 13,* 3-8.

Roy, E.A., Elliott, D., Dewey, D., & Square-Storer, P. (1990). Impairments to praxis and sequencing in adult and developmental disorders. In C. Bard, M. Fleury, & L. Hay (Eds.), *Development of eye-hand coordination across the life span* (pp. 358-379). Columbia, SC: University of South Carolina Press.

Schneck, C.M., & Henderson, A. (1990). Descriptive analysis of the developmental progression of grip position for pencil and crayon control in nondysfunctional children. *American Journal of Occupational Therapy, 44*(10), 893-900.

Schrock, R.E. (1978). Research relating vision and learning. In R.M. Wold (Ed.), *Vision: Its impact on learning* (pp. 29-50). Seattle: Special Child Publications.

Sewell Early Education Program. (1985). *S.E.E.D. developmental profiles.* Denver: Sewell Rehabilitation Center.

Starkes, J.L. (1990). Eye-hand coordination in experts: From athletes to microsurgeons. In C. Bard, M. Fleury, & L., Hay (Eds.), *Development of eye-hand coordination across the life span* (pp. 309-326). Columbia, SC: University of South Carolina Press.

Twitchell, T.E. (1965). The automatic grasping responses of infants. *Neuropsychologia, 3,* 247-259.

Twitchell, T.E. (1970). Reflex mechanisms and the development of prehension. In K. Connolly (Ed.), *Mechanisms of motor skill development* (pp. 25-38). London: Academic Press.

Uzgiris, I.C., & Hunt, J.M. (1975). *Assessment in infancy: Ordinal scales of psychological development.* Urbana: University of Illinois Press.

Vandenberg, B., & Kielhofner, G. (1982). Play in evolution, culture, and individual adaptation: Implications for therapy. *American Journal of Occupational Therapy, 36*(1), 20-28.

White, H.G. (1973). Gross motor behavior patterns in children. In C.B. Corbin (Ed.), *A textbook of motor development* (pp. 111-150). Dubuque, IA: William C. Brown.

White, R.L. (1969). The initial coordination of sensorimotor schemas in human infants: Piaget's ideas and the role of experience. In D. Elkind & J.H. Flavell (Eds.), *Studies in cognitive development* (pp. 237-255). New York: Oxford University Press.

White, R.L., Castle, P., & Held, R. (1964). Observations on the development of visually directed reaching. *Child Development, 35,* 349-364.

Williams, H.G. (1973). *Perceptual and motor development.* Englewood Cliffs, NJ: Prentice-Hall.

Williamson, G.G. (1987). *Children with spina bifida.* Baltimore: Paul H. Brookes.

**Table 6. Chad: Eye-Hand Coordination Home Program**

| Primary System Components | Developmentally Targeted Interventions | Functionally Appropriate Adaptations |
|---|---|---|
| Cognitive motivation | Sensorimotor period: object permanence, means–end, operational causality; practice games, independent solitary play, interest in material and space management | Adapted grasp switch and inclined stand for remote control of toy car; head stable in sidelying for dissociated eye movements. |
| ↓ | | |
| Visual exploration | Ocular pursuit, linked with sound and movement | |
| Cognitive motivation | Concrete operations period: collective games with rules, cooperative play | Eye-gaze board to select adapted Go Fish cards (1/4-page size, 3/8" thick lines), travel chair with head support, trunk support vest |
| ↓ | | |
| Visual exploration | Awareness of near and middle space and central field; linkage with sound and movement (mother's voice and hand) | |

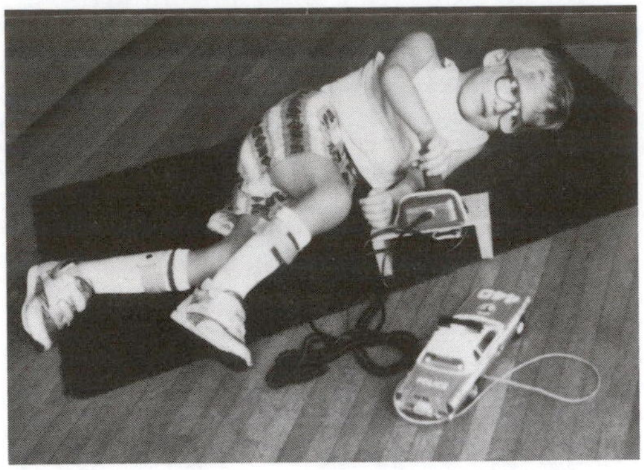

Figure 3. Toy car

Figures 4 and 5. Card game

**Table 6. Chad: Eye-Hand Coordination Home Program, continued**

| Primary System Components | Developmentally Targeted Interventions | Functionally Appropriate Adaptations |
|---|---|---|
| Tactile exploration | Accommodation (hand takes shape of object) | Head and trunk supported by mother as she assists his open hand to rub lotion on her leg |

Figure 6. Applying lotion

| Primary System Components | Developmentally Targeted Interventions | Functionally Appropriate Adaptations |
|---|---|---|
| Visual exploration ↓ | Localization in middle and far space, central and peripheral fields; identifying object by contour only | Adapted power chair with head support, tray with central joystick and dowel for left-hand stability, color-coded switches (yellow: up; green: down; blue: start/stop; red: alarm), e.g., pushing green moves the chair down for reaching fire hydrant |
| Tactile exploration ↓ | Approach, grasp, and manipulation (experiencing size, shape, texture, and accommodating hand to shape of object) | |
| Eye–hand interaction | Visually guided exploration, repetition, adaptation | |

Figures 7, 8, and 9. Exploring the neighborhood

**Table 6. Chad: Eye-Hand Coordination Home Program, continued**

| Primary System Components | Developmentally Targeted Interventions | Functionally Appropriate Adaptations |
| --- | --- | --- |
| Tactile exploration | Experiencing texture and temperature (water), size and shape (sponge) | Prone stander at kitchen sink (trunk support for arm mobility); hand and object both in visual field |
| ↓ | | |
| Eye–hand interaction | Approach, grasp, and visually directed manipulation (scrubbing with sponge) | |
| Cognitive motivation | Preoperational period: dramatic, associative, symbolic play; interest in material management | Four-choice board on lazy susan to select objects with velcro (doll's bottle, toothbrush and toothpaste) |
| ↓ | | |
| Eye–hand interaction | Visually directed reaching, grasping, and releasing into father's hand | |

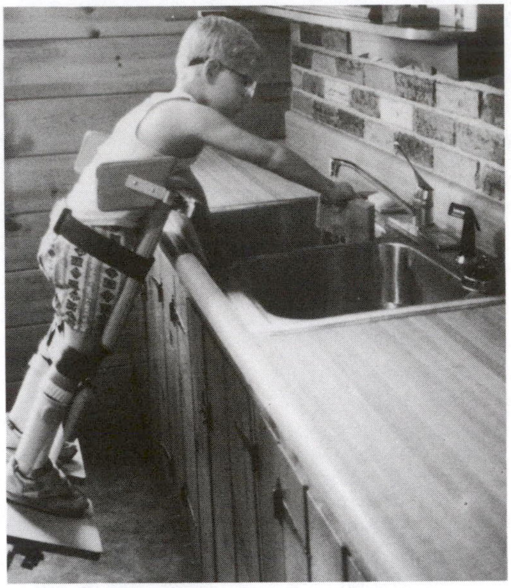

Figure 10. Water play

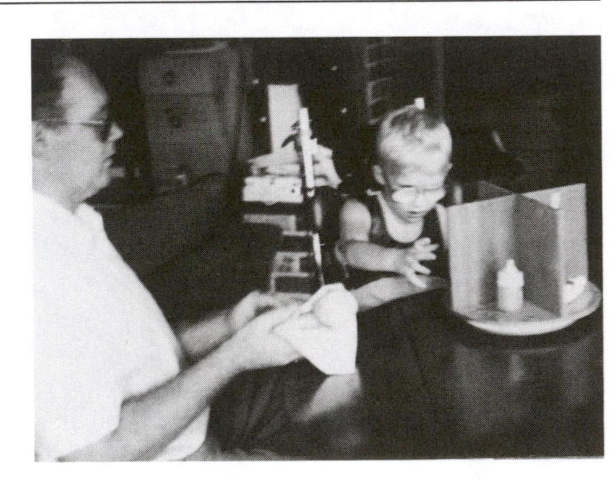

Figure 11. Doll play

**Table 6. Chad: Eye-Hand Coordination Home Program, continued**

| Primary System Components | Developmentally Targeted Interventions | Functionally Appropriate Adaptations |
|---|---|---|
| Cognitive motivation<br>↓ | Sensorimotor period: spatial relationships, operational causality; preoperational period: construction; independent play; exploration of space and material management | Power chair with velcro tray and blocks, neoprene thumb splint, dowel for left-hand stability |
| Eye–hand interaction | Reaching, grasping, and releasing velcro blocks on tray (assistive release) | |

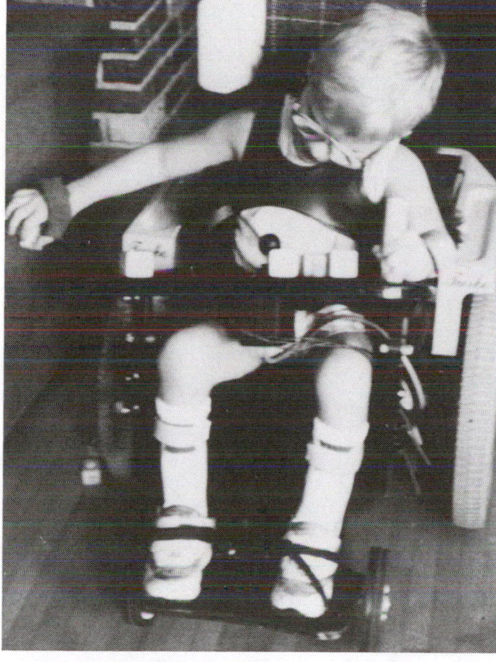

Figures 12 and 13. Block play

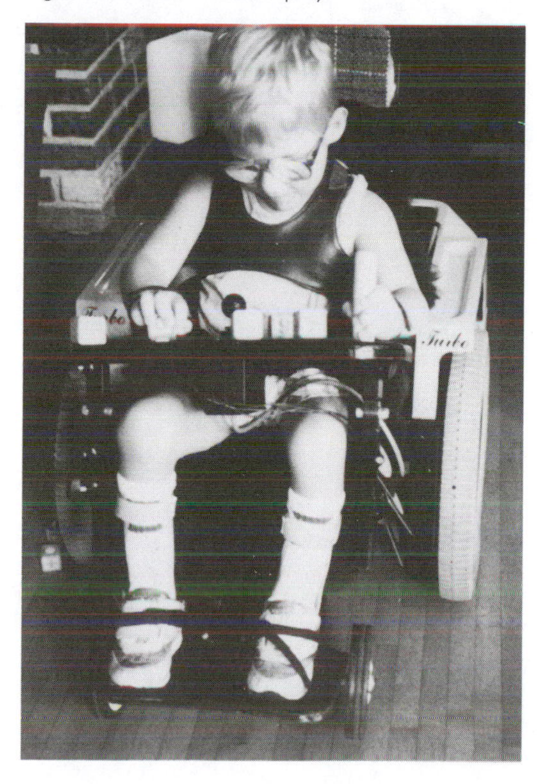

**Table 6. Chad: Eye-Hand Coordination Home Program, continued**

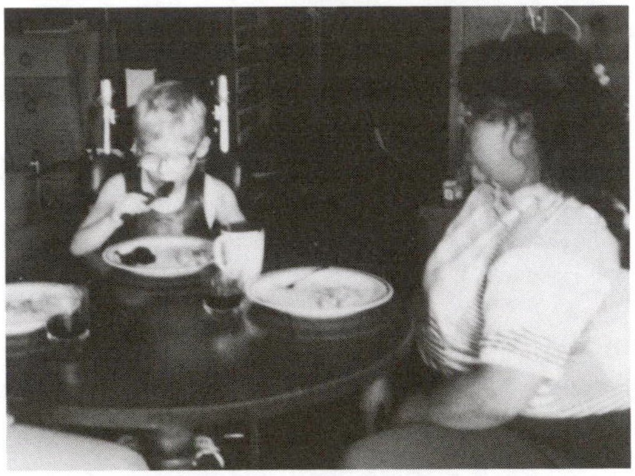

| Primary System Components | Developmentally Targeted Interventions | Functionally Appropriate Adaptations |
|---|---|---|
| Visual (and olfactory) exploration | Localization and gaze shift in near and middle space and central field; linkage with food odors | Travel chair with head support, trunk support vest, elevated table surface for arm support, adapted spoon |
| ↓ | | |
| Tactile and gustatory exploration | Reaching and manipulation (finger-feeding pudding to experience texture and taste) | |
| ↓ | | |
| Eye–hand interaction | Visually directed scooping and spoon feeding | |
| ↓ | | |
| Purposeful activity | Self-help activity; tool use | |
| Cognitive motivation | Sensorimotor period: spatial relationships; preoperational period: construction; imitation of material and space | Vertical standing frame, slant chalk board/ easel, large chalk, dowel for left-hand stability; hand and object both in visual field |
| ↓ | | |
| Eye–hand interaction | Visually directed drawing | |
| ↓ | | |
| Purposeful activity | Goal-directed use; developmental drawing sequences (vertical and horizontal lines) | |

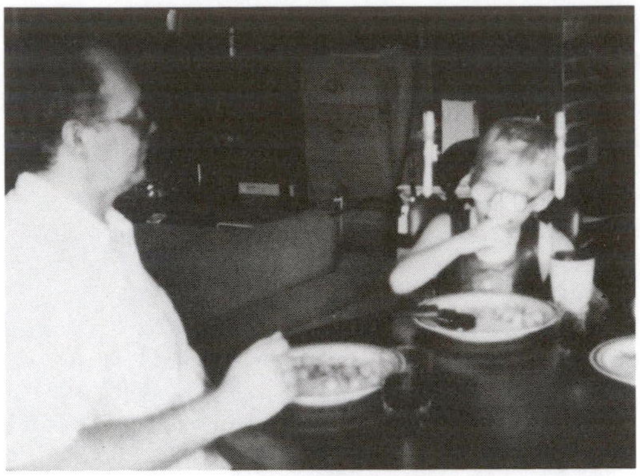

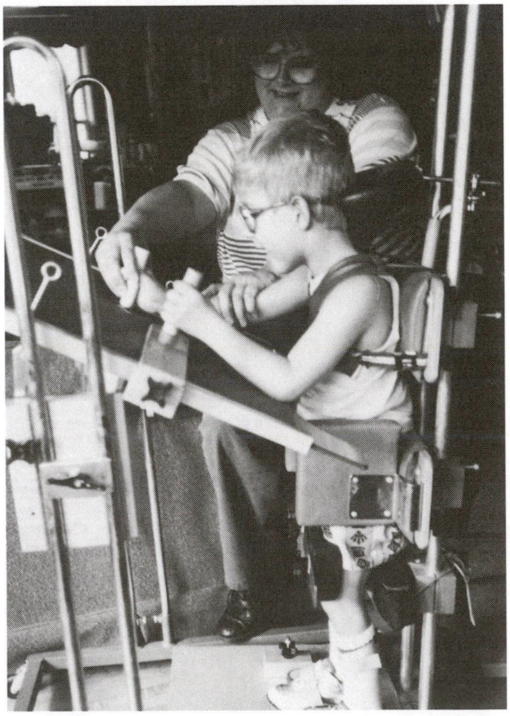

Figures 14 and 15. Feeding

Figure 16. Prewriting activities

**Table 6. Chad: Eye-Hand Coordination Home Program, continued**

| Primary System Components | Developmentally Targeted Interventions | Functionally Appropriate Adaptations |
|---|---|---|
| Cognitive motivation ↓ | Sensorimotor period: operational causality; preoperational period: imaginative play; interest in material management | Prone on floor |
| Visual exploration ↓ | Gaze shift in near space and central field; linkage with sound (talking book); developing eye–head control in downward gaze and shoulder stability through weight bearing | |
| Purposeful activity | Turns pages with one hand while weight bearing on other forearm | Adapted dimmer switch to turn talking book on/off |
| Cognitive motivation ↓ | Sensorimotor period: means–end; preoperational period: imaginative, independent play; imitation of material management | Travel chair with head support, large screen TV, large colored stickers on TV buttons corresponding to colors on power chair switches (yellow: sound up; green: sound down; blue: on/off; red: change channel) |
| Visual exploration ↓ | Awareness of near space and central field; linkage with sound, light, and movement | |
| Purposeful activity | Goal-directed operation of television | |
| Cognitive motivation ↓ | Sensorimotor period: operational causality, scheme action; preoperational period: symbolic games; interest in space and material management | Adapted power chair with head support, thumb splints, toggle treadle switch to operate LED (light-emitting diodes) scanning system developed by local electrical engineering university students, using large print materials (20/200 picture cards) |
| Visual exploration ↓ | Awareness of near and middle space (16–22") and central field; linkage with light and movement | |
| Purposeful activity | Goal-directed preacademics | |

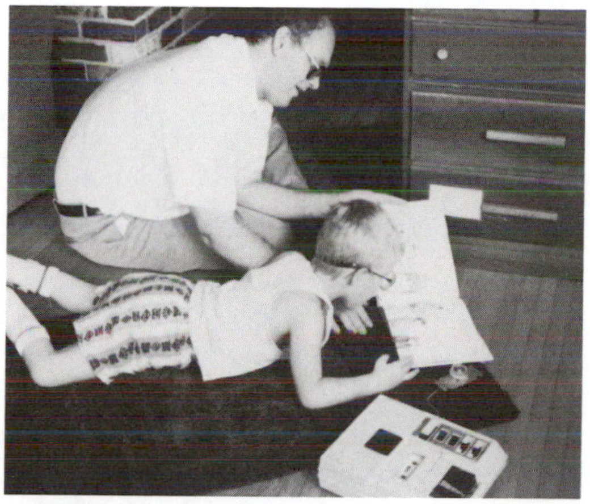

Figure 17. Reading

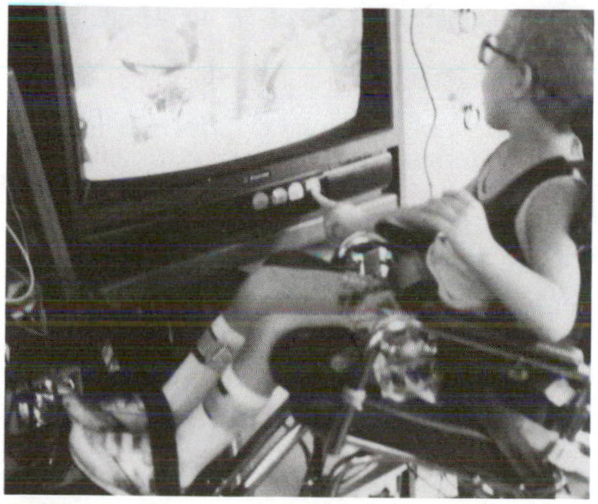

Figure 18. Television

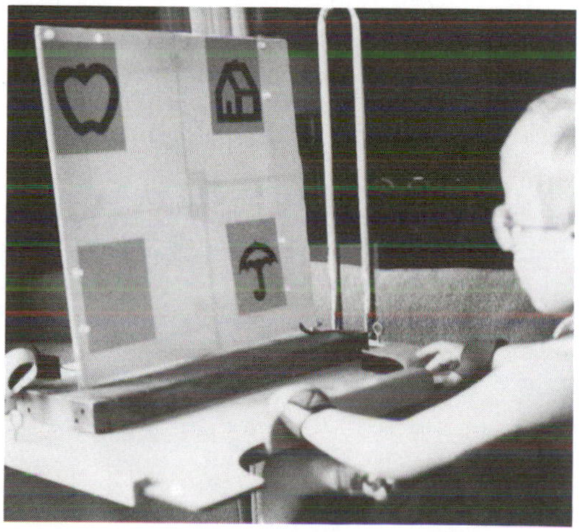

Figure 19. Augmentative and alternative communication

# 3

# In-Hand Manipulation Skills

*Charlotte E. Exner*

Young children's hand function, like other areas of development, falls along a continuum in terms of proficiency. Some children are quite competent and are considered to have excellent fine-motor skills. Most children are described by their parents and teachers as having average fine-motor skills, which typically means that they are doing fine-motor tasks with proficiency similar to a majority of their peers. A few children are identified by parents, teachers, or other individuals who spend time with them as being clumsy with fine-motor tasks.

Clumsy children are often able to perform the basic skills of reach, grasp, and release of objects effectively, but have difficulty manipulating materials. They typically have problems with acquisition of tool-use skills (e.g., using crayons, pencils, and scissors), dressing (e.g., buttoning and tying), mature independent eating skills (e.g., using a knife; opening small packages and containers), constructive play (e.g., building with blocks and other objects), and tabletop games (e.g., handling dice, moving markers, managing play money). They may continue to use two hands for manipulation of materials when one hand would be more efficient, and may drop materials often. Some children resort to pushing and pulling materials with force when they have difficulty with manipulation, which can result in materials being broken or crushed.

Impaired or delayed manipulative skills lead to inefficiency in task completion, such that children are slow and awkward or are unable to complete the task. These problems can and often do lead to the child's avoidance of activities that require object manipulation. During the preschool years and early school years, manipulation difficulties can have a significant negative influence on the child's initiative with fine-motor tasks and feelings of competence, and they often have a negative impact on the child's social acceptability to other children and adults.

Despite the attention that has been given to difficulty with fine-motor skills by professionals who work with young children, very little has been written that describes how clumsy children differ from normal children in actual performance of fine-motor tasks. Also, while types of grasp patterns and development of the opposed grasp have been described in detail in the literature, differentiation of various manipulation skills has received little attention. In most cases, the term *manipulation* appears to be used to refer to general movement of an object in space or in conjunction with other objects. However, it may also be used to refer to *in-hand manipulation* (Exner, 1989). In-hand manipulation is the process of using one hand to adjust an object for more effective object placement in that hand prior to use, placement, or release; the object remains in that hand and usually does not come in contact with a surface during in-hand manipulation.

This chapter presents an overview of approaches for viewing fine-motor-skill development. This overview is followed by a review of the literature on manipulation skills, including definitions of manipulation, muscle use during manipulation, and development of manipulation skills. Also, a recent study that addressed construct validity of a test of in-hand manipulation skills is reported.

## Overview of Fine-Motor Skill Development

Several approaches to the study and description of hand function development are reported in the literature. Regardless of the approach, however, emphasis is placed on the 1st year of life. Halverson (1943, cited by Connolly & Elliott, 1972) stated that the major developments in grasp occur during this time period, because by 14 months of age the baby can use a pincer grasp with essentially the same skill as an adult. In describing hand function development, most authors address reach and bilateral skills but primarily emphasize development of grasp skills.

Corbetta and Mounoud (1990) describe the sequence of development for unilateral skills and bilateral skills. They state that unilateral and bilateral skills develop sequentially rather than simultaneously, in that unilateral skills precede bilateral skills in their level of complexity. Unilateral skills typically involve only one object, although this object can be used with another object not being held. Bilateral hand skills (not reaching) are believed to be more difficult and are usually not seen until close to 12 months of age, because the child will need to perceive more object properties and coordinate more actions.

Corbetta and Mounoud (1990) primarily address skills from birth to 24 months of age. They divide this age span into three key age groups, with the time from birth to 6 months forming one time period, 5 to 6 months to 12 months the next period, and 9 to 12 to 24 months the third. During the first phase, the primary area of development is gradual control of upper-extremity movement based on recognition of where an object is in space. In the next phase, the baby learns to accommodate the opening and shaping of the hand in response to stimuli about objects' particular characteristics. Between 9 and 24 months, the baby acquires increasing understanding about how to relate objects to one another. By the end of this period, the two hands can work independently of one another yet together in an activity that requires differentiation of hand function.

Gilfoyle, Grady, and Moore (1981) proposed dividing the development of upper-extremity/fine-motor skills into three key phases that do not have ages of skill acquisition. However, based on the task descriptions included, the last of these phases extends well into the preschool years. The three phases they described are the primitive phase, the transitional phase, and the mature phase. During the primitive phase, the upper-extremity and hand reflexes affect reach and grasp; therefore, although some motor skills begin to be demonstrated, the baby in this phase will primarily use visual interaction with objects. The transitional phase is believed to be a time of developing purposeful hand use. Mouthing, banging, bilateral reaching, and pronation and supination to move an object in space are used. During the mature phase, eye-hand coordination develops fully. Hand orientation and placement during reaching improve, and the skills of voluntary release and the pincer grasp mature. Later in this phase of development, tool use is seen; these tool-use skills mature over a period of several years.

Few studies have been conducted on the development of grasp of objects, other than pencils, after the first 1 to 2 years of life. However, Connolly (1973) studied 40 normal children during grasp of a variety of objects. He found that the 3-year-olds used grasps that required power and those that required precision on appropriate objects. With more difficult tasks, however, the children tended to revert to use of less mature patterns. He concluded that children gradually limit the variety of grasps used for particular objects to those that are the most dependable. Limiting variety within these simple patterns allows the child to focus on developing complex movement patterns. Cognitive changes, more than physical changes, seem to precipitate this change in how basic skills and movement patterns are used (Connolly).

## Definitions of Manipulation

The term *manipulation* has been used by many researchers to describe actions produced by the hands in relationship to objects. However, some researchers have used the term to refer to the movement of objects regardless of how the movement is produced, while others have been specific that manipulation involves actions on objects that are produced by the fingers. Some have indicated that manipulation is any action produced by the fingers, while others have indicated that only specific types of finger actions are true manipulation. Regardless of the way in which the term manipulation is used, the importance of this skill in human functioning is stressed. For example, Gilfoyle, Grady, and Moore (1981) state that "the evolution of manipulative prehension contributes to the uniqueness of man" (p. 152).

The following definitions illustrate the variety of ways in which the term manipulation is used. Connolly (1973) points out an aspect of development, but a nonmotor one, that may be inferred in the other authors' works and that will affect the use of manipulation skills by children. He stresses that manipulation occurs primarily as a result of schemata, which are cognitive in nature, rather than primarily as a result of motoric functions of the hand. The particular manipulative strategies used are dependent on the schemata available. "Manipulation itself involves orientation to an object" (Connolly, p. 347). Praxis is mentioned by Connolly in a manner that implies that manipulation is equivalent to practic ability.

Corbetta and Mounoud (1990) use the term manipulation throughout their chapter on development of hand functions to refer in a general way to hand functions that involve control of objects. They use the terms "one-handed object manipulation" and "two-handed object manipulation" (p. 197) to describe all types of actions produced by the hands including grasp, voluntary release, and bilateral skills. In these descriptions, objects may be used by themselves or in relationship to other objects. Corbetta and Mounoud state that "in manipulative behavior action calls for organization and adaptability of digital, manual, and arm gestures according to the physical and spatial properties of objects" (p. 189). Thus, it seems that arm, hand, and finger actions must occur together for manipulation to be present. Clearly, manipulation is a skill that only occurs in relation to objects. Corbetta and Mounoud indicate that when bilateral manipulation is used, such that both hands are separately handling objects, a wide variety of skills can be accomplished.

Gilfoyle et al. (1981) refer to "manipulative prehension activities" (p. 149). Despite the use of the term "prehension," which implies grasp, they include activities such as piano playing as well as shoe tying. Piano playing obviously requires independent, coordinated finger activity but does not have prehension as an element. These authors also included scissor and crayon use under the category of manipulation. Levine, Carey, Crocker, and Gross (1983) similarly include use of tools, such as eating utensils, pencils/pens, and craft tools, in their definition of manipulation, but they use the term "manual dexterity" to define the finger actions that are used to control tools.

Moss and Hogg (1981) also include prehension in their definition of manipulation by saying that prehension is one component of this skill. They propose that "the observation and classification of grip is a useful avenue to the study of manipulation in general" (p. 36). In assessing grip, information can be gathered about use of appropriate grasp patterns and the individual's knowledge of how the hands need to be used to perform the task effectively. For example, Moss and Hogg (1981) assessed children's grasp patterns during placement of pegs into holes when the pegboards were oriented horizontally and vertically.

## General Categories of Manipulation Skills

### Classification Systems

The grasp patterns that allow for manipulation have been identified within the descriptions of the major types of grasp. Napier (1956) differentiates between prehensile and nonprehensile movements of the hand. The nonprehensile movements are those that do not involve grasp; the movements may be whole-hand ones or isolated finger actions. Within the prehensile category, he makes a distinction between those grasp patterns used for power functions and those used for precision. He indicates that the terms could be used to indicate a holding (static) pattern or to refer to the actions involved in assuming the grasps. A term that Napier (1956) uses but does not define is *precision handling*; yet, in his use of this term, he seems to be referring to a type of hand use somewhat different from the precision grip, which he clearly defines as object contact with finger use and an opposed thumb.

Landsmeer (1962) builds on Napier's work by attempting to clarify and further define the term precision handling. Landsmeer believes that there are important differences between power and precision grips not only in configuration of hand use but also in terms of static and dynamic qualities. He states that the power grip is essentially static but the precision grip is not. "In the precision grip there is no question of a static phase, since the fingers themselves manipulate the object, and there is no point in distinguishing dynamic and static phases in this movement pattern" (Landsmeer, p. 165). Because he disagrees with the term *grip* for this dynamic movement, he suggests that the term precision handling be used. In defining precision handling, Landsmeer indicates that the object is first held by the fingers with the thumb in opposition; this positioning of the fingers on the object is then changed through small, independent movements of the finger joints. Thus the terms precision handling and manipulation are used to mean the same type of movement. Long, Conrad, Hall, and Furler (1970) elaborate on this definition of precision handling by stating that via this process the object moves "either in space or about its own axes" (p. 854).

Elliott and Connolly (1984) use somewhat different terminology to describe various types of hand movements that are used to manipulate objects. When objects are moved in space by a movement of the arm and hand together, the movements are considered to be "extrinsic" (p. 284). Manipulation of objects held in power grips will occur via these less refined upper-extremity and trunk movements. In contrast, when the fingers must act together in a coordinated manner to move the object while it stays in the hand itself, the movements are considered to be "intrinsic" (p. 284). A wide variety of movements are possible when using a finger grasp rather than a power grasp. The finger movements produce fine manipulation of an object; however, in doing so, the stability of the object in the hand and the power that can be used on or with the object is diminished (Elliott & Connolly).

Although manipulation is an important skill, various types of manipulation have only been described by Long et al. (1970), Elliott and Connolly (1984), and Exner (1989, 1990a). However, Parker and Gibson (1977) do describe types of object manipulation and tool use in cebus monkeys and great apes. They use five categories of manipulation, which progress from unilateral or bilateral object grasp, through simple object manipulation, object-substrate manipulation (use of an object in relationship to a surface), complex object manipulation, and social-object manipulation. Their definitions indicate a range of skills that they consider to be related to cognitive skills present in these animals. These movements, however, correspond more to the type of manipulation described by Elliott and Connolly as extrinsic movements. The terms are used by Parker and Gibson to mean not only the complexity of the action but also the complexity of the goal for which the actions are being used.

Long et al. (1970) uses the terms *precision translation* and *precision rotation* within the general category of precision handling. Precision translation involves using the fingers and thumb to move an object between the finger surface and the palm, such that at one point the object is in contact with the palm and at another point it is positioned between the thumb and fingers; with movement, the location of the object changes within the hand. An example given to illustrate precision translation is using the fingers to put thread through a needle or pulling thread after it is inserted in the needle's eye. The term precision rotation is used to describe "thumb and finger activity designed to rotate an object about one of its internal axes" (Long et al., p. 854). These authors indicate that pinch is not precision handling because it is a static pattern.

### Muscles Used During Object Manipulation

Long et al. (1970) studied two aspects of precision handling using electromyographic (EMG) recordings of muscle activity. For assessment of translation activities toward and away from the palm, they used a spring-type device that had a piece that could be pulled or pushed. For rotation, they used a rotor positioned horizontally to the surface; different sizes of shafts were used. The power required to operate either of these devices could be varied by modifications to them. Long et al. (1970) tested 10 normal adults and found that the type of muscle activity used in precision handling was quite consistent from one adult to another.

They did find some variations in muscle activity use depending on whether force was required. During translation toward the palm, they documented primarily extrinsic hand muscle activity, with some activity in opponens pollices and abductor pollices brevis. In contrast, when force was used for translation toward the palm, the interossei (not the lumbricales) were active, as well as the flexor pollicus brevis, opponens pollices, and abductor pollices brevis ("the thenar triad") (Long et al., p. 864) and adductor pollices. For translation away from the palm when force was not used, the lumbricales were active; a small amount of activity was also noted in abductor pollicus brevis and opponens pollices. When force was required in this pattern, the thenar triad was involved as well as the interossei and the lumbricales. The primary muscles found to be active in precision rotation were the interossei, which are important in finger abduction and adduction, as well as the thenar triad. The lumbricales were noted to function differently with different types of rotation. Obviously the intrinsics, which include all the interossei, the lumbricales, the thenar triad muscles, and adductor pollices, are very important in the use of manipulative skills. Understanding the different patterns of muscle activity in the different types of manipulation could be useful if a pattern of specific muscle weakness was present.

## Detailed Descriptions of Manipulation Skills

### Elliott and Connolly's Classification System

Elliott and Connolly (1984) provide a very comprehensive description of various intrinsic hand movements: They attempt to describe all the common intrinsic manipulative patterns shown by normal individuals (p. 284). Intrinsic movements can cause linear object movement or object rotation. Their classification system employs two broad groups of patterns: simultaneous and sequential. The simultaneous patterns usually are characterized by one pattern of finger movement that can be reversed. In contrast, the sequential patterns involve movement of the object with some fingers while others stabilize the object, then a repositioning of the fingers such that the fingers and thumb alternate between stabilizing and moving the object. The object is moved by a sequential action of the fingers and thumb.

The simultaneous patterns include both simple and reciprocal synergies (Elliott & Connolly, 1984). In simple synergies, the thumb and fingers act together to move into a flexion pattern. Examples of simple synergies include the pinch grasp, the dynamic tripod, and a squeezing action (Elliott & Connolly). Reciprocal synergies are those patterns in which the thumb and fingers move in opposite directions; the thumb and fingers move independently of one another. The thumb may move in either a flexion/extension or an abduction/adduction pattern. Terms used to describe reciprocal synergies in which the thumb moves in abduction/adduction are "twiddle" and "rock;" these patterns are both used to rotate an object. With the thumb moving in flexion/extension, the patterns of "radial roll," "index roll," and "full roll" are seen. The first two thumb flexion/extension patterns involve primarily the thumb with the index finger, while the full roll includes other fingers as well.

Three key sequential patterns were described by Elliott and Connolly. These patterns all involve changes in object position on the finger pads during the movement. They include the "rotary step," the "interdigital step," and the "linear step." In both the rotary step and the interdigital step, the object rotates, but in the linear step the object is slid across the pads of the fingers.

Finally, Elliott and Connolly (1984) describe a pattern in which part of an object is stabilized in the hand while the index finger and thumb are involved in manipulating part of the object. Examples are tying, and pushing a top off a pen while stabilizing the pen.

Within each of the categories of skills described, variations occur. The names of the patterns usually vary as a result of more fingers being added to a pattern. The pattern of manipulation may be varied under situations of demands for different degrees of force or amounts of speed, or due to the object characteristics (Elliott & Connolly, 1984).

This system classifying manipulative movements has some advantages, but it also has several disadvantages. It is a valuable classification process for studying manipulative skills in normal adults because Elliott and Connolly (1984) have presented such a thorough description of each pattern. Research studies may be enhanced by the use of this level of analysis of manipulation skills. However, this system appears to be difficult to use clinically due to the complexity of the classification system and the language used to describe patterns. There are many skill categories and, for several skill patterns, distinguishing between thumb adduction/abduction and thumb flexion/extension is required. In normal individuals, these movements occur very quickly and may be difficult to discriminate. Also, because these skills are at a very high level, it is likely that many people with hand-function problems would be unable to do most of the skills. Another difficulty with this classification system is that it does not address handling several objects at one time within the hand or moving objects into and out of the palm. All of the actions described happen solely on the finger surface, with the exception of the "combined" pattern. Finally, the terminology used is not currently familiar to therapists and is not easily adapted for communication with parents and teachers; thus, communication among therapists and others would not appear to be facilitated by use of this system.

## Exner's Classification System

Because manipulative skills seem to be critical in describing children who are clumsy and differentiating them from children with fine-motor skills within normal limits, I have been attempting to define the construct of in-hand manipulation. From 1980 until 1984, I was in the process of developing a manipulation skills classification system for therapists to use in clinical settings as well as in research studies. The work by Long et al. (1970) was the primary basis for my early efforts in this area. During that time I began using the term *in-hand manipulation*, observed children with neuromotor problems for the presence/absence of these skills, and treated children for deficits in these skills. In-hand manipulation was defined as the process of adjusting objects within the hand after grasp (Exner, 1989, p. 242) This adjustment was believed to be important for optimal object orientation within the hand so that the objects could be used or released with control (Exner, 1990a). Initially, I used three broad categories of skills, two of which were adapted from Long et al. These three categories were: translation (fingers-to-palm or palm-to-fingers), shift, and rotation (Exner, 1989). I also defined these skills as occurring with or without simultaneous stabilization of other materials in the hand (Exner, 1989). More recently, I determined that rotation needed to be divided into two categories—simple and complex rotation (Exner, 1990a). The definitions that I am currently using are as follows:

### *Translation*

Translation is a linear movement of the object in the hand from the finger surface to the palm or the palm to the fingers (Long, et al., 1970). To be considered full translation, the movement must either begin or end on the fingers distal to the DIP joints (Exner, 1989).

#### ***Finger-to-Palm Translation***

In this type of translation, an object is held by the distal finger surface (the pads of one or more fingers or against the volar surface of one or more DIP joints) and the pad of the thumb and moved into the palm proximal to the metacarpal-phalangeal joints. For example, finger-to-palm translation is used when picking up a coin from a surface and moving the coin to the palm of the hand.

#### ***Palm-to-Finger Translation***

In this type of translation, objects are moved from some area of the palm out to the distal finger surface. The thumb is active in producing object movement; it moves from a pattern of flexion to extension. For example, palm-to-finger translation is used to move a coin from the palm of the hand out to the fingertips for placement into a vending machine.

### *Shift*

Shift movements occur at the finger and thumb pads with alternation of thumb and (usually radial) finger movement. Shift may be used for the final adjustment of an object after grasp or after use of another in-hand manipulation skill; in this case it is used for refinement of placement of the object against the fingers and thumb. An example of shift is moving a coin that is positioned near the DIP joints of the fingers further out onto the pads of the same fingers or adjusting a pen in the hand so that the fingers and thumb are closer to the writing end of the pen. Shift also may be seen in movement of an object across the pads of the fingers, usually in an ulnar-to-radial direction. This occurs when an object is held with the ulnar fingers and the thumb but needs to be moved to the radial fingers for use or placement. Fanning several playing cards held in one hand may require this type of shift.

### *Rotation*

Rotation is movement of an object around one or more its axes (Long et al., 1970). These movements occur at or near the pads of the fingers.

#### ***Simple Rotation***

Simple rotation occurs when an object is turned or rolled between the pads of the fingers and the pad of the thumb via an alternation between thumb movement and finger movement. It may occur with only the index finger and thumb or with involvement of additional fingers. When more fingers are used, they typically act as a unit. Finger adduction is typically present. An example of simple rotation is rolling a small ball of clay back and forth to form it into an elongated shape or unscrewing a jar lid. The object is usually not rolled over completely, but instead is rotated for about one-fourth or one-half its circumference.

#### ***Complex Rotation***

Complex rotation involves rotation of the object or rotation of the object that requires isolated, independent movements of the fingers and/or thumb. The object is turned between 180 and 360 degrees. The object is typically stabilized alternately by the fingers and thumb. When the object is stabilized by the thumb, it is moved by the fingers; when it is stabilized by the fingers, the thumb moves the object. The fingers often alternate during complex object rotation, with the radial fingers stabilizing and the ulnar fingers moving the object and then the radial fingers moving the object. An example of this activity is seen in turning a paper clip so that the opposite end can be used for placement on a piece of paper.

Each of the above in-hand manipulation skills may occur as described or while the individual is simultaneously stabilizing one or more other objects in the same hand. The ulnar fingers stabilize the object while the radial fingers perform the in-hand manipulation. The stabilized object is usually held in the central or ulnar part of the palm. When stabilization of objects within the hand occurs during in-hand manipulation, the skill is named and the phrase "with stabilization" is added. For example, "simple rotation with stabilization," "finger to palm translation with stabilization," and so forth, would be used. An example of palm-to-finger translation with palmar stabilization is holding several coins in one hand and using palm-to-finger translation to move only one coin to the pads of the fingers while retaining grasp of the other coins.

## Development of Manipulation Skills

Despite the apparent significance of in-hand manipulation skills to the performance of many daily life tasks, little study of the development of these skills has been done. However, several authors have made general statements about the development of manipulation skills. Some of these statements are confounded by the differences in the meaning that these authors have given to manipulation skills. For example, if they use the term *manipulation* to mean any handling of objects, they are likely to indicate that manipulation begins early in the 1st year of life. The statement by Corbetta and Mounoud (1990) that "an infant's ability to manipulate [begins] the moment he is able to grasp, orient, or adjust his hand to objects based on tactile, visual, or auditive information, that is, at around 5 to 6 months" (p. 196) illustrates this view of manipulation development. However, these authors distinguish between extrinsic manual activities (p. 199), which are more gross upper-extremity movements with objects, and actions that involve specific finger activity. The latter skills are described as beginning prior to 1 year of age.

### Object Manipulation in Infants

Ruff (1984) investigated the age at which infants use specific finger movements in exploring objects. She used a process of familiarization trials with infants between the ages of 6 and 12 months in order to determine the influences of different colors (same texture) and different shapes (same visual pattern) on the infants' manipulation strategies. She found that both duration and frequency of mouthing decreased with age while fingering of objects increased with age. Rotary movement of the wrist (probably forearm supination and pronation) was used similarly by infants of all ages, but rotation of the object in the process of moving it with two hands clearly increased during this age range. Ruff (1984) also documented increased fingering with the textured objects. She suggested that during fingering, the infant receives visual and tactile information simultaneously. On the other hand, during bilateral manipulation the object can be moved in many ways, and all parts of the object can be viewed. This type of manipulation allows the infant to obtain a great deal of information about the object's size, shape, and other structural characteristics. Infants typically modify the skills used in order to gain more information about objects, and their motor skill development allows them to change the skills used (Ruff, 1984).

Ruff (1984) followed this study with another one in which she varied shape, texture, and weight of objects in familiarization trials presented to 48 children aged 9 and 12 months. She found that with texture changes the infants showed increased fingering of the surface. With shape changes, hand-to-hand transfer of the objects increased in both groups of infants; rotation increased significantly with shape change in the 12-month-olds only. They did not show a change in behavior with the weight changes. Thus, it appears as if even young infants can systematically vary their use of manipulation skills to obtain information about specific object characteristics.

## In-Hand Manipulation Development in Young Children

In an attempt to determine the ability of specific tasks to elicit in-hand manipulation skills, the general age range during which these skills develop, and a tentative sequence of these skills, I conducted a study of normal children's manipulation skills with a variety of materials (Exner, 1990a). The children were between 18 months and 6 years, 11 months of age. Spontaneous in-hand manipulation skills and substitution patterns used to complete the tasks were recorded. Key findings regarding these in-hand manipulation skills were:

- Between 18 months and 7 years of age, the variety of in-hand manipulation skills increased such that all skill types were present by the end of this age range. However, the oldest children in the study did not have adult proficiency in use of the skills and tended to revert to less mature patterns at times. The use of more complex manipulation skills may be significantly affected by cognitive development and need for tool use.
- The age period during which the greatest number of skills seemed to emerge in these children was between 2 and 2 1/2 years.
- Most of the in-hand manipulation without stabilization skills were seen in children by the age of 3 years.
- No stabilization of materials during in-hand manipulation was observed in children under age 2.
- New skills were usually seen first with small-sized objects rather than medium-sized or tiny objects. Object size definitions are relative to the child's hand size, but generally medium size is 1 inch in all dimensions or greater than 1 inch in one dimension; small is usually greater than 1/2 inch but no more than 1 inch in any dimension; and tiny is usually 1/2 inch or less in all dimensions.

A subsequent study was designed to assess the differential effects of cues on young children's use of in-hand manipulation skills (Exner, 1990b). In this study, 12 3-year-olds and 16 4-year-olds with no known developmental problems were given 26 in-hand manipulation items without any cues as to how to perform the task. They were only told what the task was (e.g., "put the key in the lock"). Their spontaneous in-hand manipulation skills were recorded. Within 1 week of this test, each child was given the test again, either with verbal cues for or demonstrations of the in-hand manipulation skills to use in completing each item. For example, children in one group were told to try to perform the skill without using the other hand or table (a verbal cue). The other children were told to observe the principal investigator and try to perform the skill in the same way.

Both groups of children showed significant improvement with the cues, but there was no difference between the two groups' mean scores. However, with verbal cues, the children were more consistent across the three trials of each item presented than they were with the demonstration cues. The 4-year-old boys performed in a manner similar to the 3-year-old boys and girls; the 4-year-old girls were much more proficient in their skills. The lack of a developmental trend toward greater use of in-hand manipulation skills may have been due to the fact that the test used in this study was not designed to assess the quality of manipulation.

The two studies described above lend some support to the general statements made by several authors regarding the development of manipulative skills. Corbetta and Mounoud (1990) note that precise use of manipulation skills occurs later in development and that intrinsic hand movements are seen much later than extrinsic ones. Elliott and Connolly (1984) also make the point (based on studies that they conducted related to patterns used in handling tools) that manipulation skills improve substantially by the age of 5 years, but that young children can be quite variable in the patterns that they use. Gilfoyle et al. (1981) stress that it is through involvement in purposeful activities that manipulation skills continue to develop and that this development occurs across a span of several years. Eventually, adults can use either hand for a wide variety of manipulation skills (Connolly, 1973). However, as with other developmental skills, manipulative skills are not specifically taught to normal children (Elliott & Connolly, 1984); the skills are acquired in the course of interaction with a variety of objects, including tools (Exner, 1990a; Gilfoyle et al.).

## The Test of In-Hand Manipulation by Exner (TIME)

No standardized test for specific in-hand manipulation skills is presently available. To address this problem, I am developing the Test of In-Hand Manipulation by Exner (TIME) for use by occupational therapists in evaluating in-hand manipulation skills of preschool and young school-age children who have or are at risk for having fine-motor problems. Currently, I am in the process of assessing the test's validity and reliability. The most recent study of this test (Exner, 1991) was designed primarily for the purpose of assessing rater reliability and content validity. Therefore, a relatively large number of experienced pediatric occupational therapists (24) served as raters for a relatively small number of children's videotapes (16). Scoring of the children's performances was done by six raters for each child, and each child was rated twice by the same six raters. The generalizability coefficient for ratings made by six raters was .93, and internal reliability was high (coefficient alpha = .98). Although construct validity was not the focus of this study, tentative information about the test's construct validity and development of children's in-hand manipulation skills can be derived from the data collected.

The 8 boys and 8 girls who were videotaped for the study represented the following age groups: 2 years (n = 3), 3 years (n = 3), 4 years (n = 3), 5 years (n = 3), and 6 years (n = 4). Their overall fine-motor skills were assessed with the Peabody Fine-Motor Scale, in addition to the TIME. Four children had diagnoses that indicated developmental dyspraxia, developmental delay, or a neuromotor problem. The other 12 children were considered normal, as they had no identified developmental problems and were not receiving any special services.

The version of the TIME used in this study was based on earlier versions of the instrument that were used in pilot studies. The tasks selected for inclusion on this test were those that children have been observed to complete using in-hand manipulation skills, and most were on earlier versions of the TIME. The test requires that the child be able to follow simple directions and to understand how objects can be combined in gamelike or constructive play situations. Therefore, some young children with cognitive delays, who are chronologically 18 months of age or older, may not be able to complete many of the items on the test. However, most children who are 18 months or older and who are in the last phase of Piaget's sensorimotor stage of development ("inventions through sudden comprehension" [Gallagher & Reid, 1981, p. 74]) have been observed to be successful with task completion even if they do not use in-hand manipulation skills. For example, children who could not use palm-to-finger translation could still put money in the bank, and those without complex rotation skills could use both hands together to turn over pegs to put them into holes. All items are meant to be perceived as play items by the children. As with all motor tests, however, cooperation with the test directions affects performance.

The test version used in this study was composed of 28 activities, each of which was scored for one or more types of in-hand manipulation; a total of 101 items were scored. The score assigned for each item represented the child's best performance on the two trials administered. Items were scored with a 5-point rating scale: 0 on the scale indicated that no manipulation was present; 4 indicated that the particular manipulation skill was present and completed efficiently. In addition, substitutions for in-hand manipulation skills were recorded. As in the earlier study, children were told the end product for each task but not the process to use in accomplishing the tasks. See Table 1 for examples of the test items for various in-hand manipulation skills.

### Chronological Age and In-Hand Manipulation Skill

The data suggest that the nondisabled children's TIME scores generally increased with increased chronological age. The two children at 2.1 years had mean TIME scores of .99 (out of a possible overall mean of 4.0). The two children between 2.7 years and 3.2 years had a mean TIME score of 1.21. The two older 3-year-olds both had means of 1.69. Overall, the mean for the six children from 4.5 to 6.0 years was 2.08 (range = 1.93 to 2.28). However, these children did not show a clear upward trend in mean scores; the oldest child (6.8 years) and one of the older 4-year-olds had similar scores.

These data support findings from earlier studies that younger children clearly have less skill in in-hand manipulation, as measured by the TIME, than do the older children. One interpretation of the finding that a less clear upward trend was found in the 4 1/2 to 6-year-olds is that some of these particular children's scores may not be representative of scores that would be typical of children at these chronological ages; with such a small sample of children, it is not possible to determine which scores may be most typical versus most unusual for children of particular ages. It is also possible that the TIME, in its current version, does not identify the qualitative changes that occur between the ages of 4 1/2 and 6 years. Another explanation is that the manipulative

**Table 1. Sample Activities Used in the TIME to Test In-Hand Manipulation Skills**

| Skill | Activity |
|---|---|
| Finger-to-palm translation | Child picks up quarter and "hides" it in the same hand. |
| Finger-to-palm translation with stabilization | Child has two quarters in one hand; he or she picks up the third quarter and "hides" it in the same hand. |
| Palm-to-finger translation | Child moves one 3/4" cube from palm to fingertips for stacking. |
| Palm-to-finger translation with stabilization | Child has two 3/4" cubes in one hand; he or she moves one from palm to fingertips for stacking. |
| Shift | Child holds marker at midpoint and moves fingers downward toward writing end for placement prior to writing. |
| Shift with stabilization | Child has one key in palm of hand and moves another key from middle phalanges of radial fingers to pads of same fingers for placement in lock. |
| Simple rotation | Marker is placed horizontally on table with writing end at *ulnar* side of child's hand; child picks up and rotates marker for writing. |
| Simple rotation with stabilization | Child has two keys in palm of one hand (with holding end aligned with radial fingers) and moves one from palm to finger surface; he or she then rotates the key for placement in lock. |
| Complex rotation | Marker is placed horizontally on table with writing end at *radial* side of child's hand; child picks up and rotates marker for writing. |
| Complex rotation with stabilization | Two 1 1/2" long pegs shaped as persons are placed in child's hand with "heads" at ulnar side of child's palm; after child moves one peg to finger surface, he or she rotates peg for placement in pegboard. |

skill changes that occur within this age period may be speed related rather than quality related. All of these possibilities can be assessed in studies with a larger sample of children by first assessing for quality of their in-hand manipulation skills and then assessing for speed in task performance.

## General Fine-Motor Skill and In-Hand Manipulation Skill

When all 16 children were ranked from low to high by their Peabody Fine-Motor Scale raw scores, a similar trend of increasing skill on the TIME was noted. Children with the lowest Peabody raw scores (144 to 155; n = 4) had the lowest TIME scores (.52 to 1.19). The seven children with the highest Peabody raw scores (205 and 220) had the highest TIME scores (1.93 to 2.23). TIME scores for the five children with Peabody raw scores of 177 to 192 were in a broad range. However, when the lowest and highest TIME scores were omitted from this group, the other scores were between 1.69 and 1.74.

In comparing children's fine-motor scores as represented by their Peabody Fine-Motor Scale scores and their TIME scores, differences in the two tests must be considered. The Peabody Fine-Motor Scale contains many items that do not require fine manipulation of materials. Also, scoring on the Peabody items is primarily based on task completion, with speed of performance a consideration in scoring of some items. Quality of performance is not directly tapped. The TIME, however, is designed to assess quality of performance, and task completion is not specifically part of the scoring. That is, each TIME score is based on the method the child uses in task completion, not on whether the task is successfully completed. Speed is only one component of the TIME scoring criteria, and this judgment of speed is based on the rater's opinion, not on timing with a stopwatch. Therefore, due to the scoring differences and different purposes of these two fine-motor tests, it is expected that TIME scores will moderately correlate with Peabody Fine-Motor Scale scores. This relationship needs to be tested with a larger sample of children than included in this pilot study.

## Difficulty of In-Hand Manipulation Skills

In order to determine the difficulty of the various in-hand manipulation skills, mean scores for each of the scales were rank ordered from the highest to the lowest. As in the pilot study by Exner (1990a), the skills with stabilization all had lower mean scores (were more difficult) than the same skills without stabilization. This finding was expected, due to the differentiation of movements within the hand that must be present in order for the child to be able to hold some objects with the ulnar fingers while manipulating with the radial fingers. Finger-to-palm translation and simple rotation were found to be the easiest skills. The low mean scores for complex rotation and shift indicate that these are more difficult for children to use than are the other in-hand manipulation skills.

The difficulty order of these skills, as reflected by the mean scores, seems related to the development of skill in using various finger and thumb movements. Finger-to-palm translation and simple rotation involve much less isolated, independent thumb movement than do palm-to-finger translation and complex rotation skills. Complex rotation, which requires thumb and finger movements that are differentiated from one another, is the most difficult skill. Shift is apparently almost as difficult. Perhaps this is related to the fact that shift must occur at the finger pads and requires use of the intrinsic hand muscles in isolation from the extrinsic hand muscles to produce well-controlled metacarpal-phalangeal flexion and interphalangeal extension.

In addition to the effects of finger movement control on in-hand manipulation skill difficulty, typical sequences of skill use within activities may influence the ease and frequency with which a particular skill is used. Although it appears that simple rotation is slightly easier than palm-to-finger translation, simple rotation with stabilization appears to be slightly more difficult than palm-to-finger with stabilization. This difference may be due to the fact that palm-to-finger with stabilization always precedes simple rotation with stabilization within the tasks on this test. (For example, when testing simple rotation with a peg, the child first held three pegs in the hand, then was asked to bring one peg out of the hand to put it in a pegboard. In this process both palm-to-finger translation and simple rotation were tested while the child was holding two other objects in the hand.) In contrast, simple rotation and palm-finger movements are not always tested together.

## Relationships of In-Hand Manipulation Skills to Total TIME Score

Correlations of the skills with one another and with the total test scores were used to determine the relationship of each skill to overall in-hand manipulation skill. With the exception of the shift with stabilization and the complex rotation with stabilization scales, the scale-total correlations fell between .55 and .88. The high and moderate relationships of palm-to-finger with stabilization and simple rotation with all or most of the other scales and their high relationships with the overall test score suggest that these two skills are critical in the assessment of in-hand manipulation skills. In contrast, the much lower correlations of the shift with stabilization and the complex rotation with stabilization scales with the other scales and with the total score suggest that these scales may either be important and tapping very different skills than the other scales, or that they have too few items to contribute effectively to the total test score.

## Summary

Several theories have been proposed to explain general fine-motor development, grasp skill development, and development of bilateral skills. Most of these approaches refer to manipulative skills, but only by indicating that these skills are at the highest level within the general area of fine-motor skill development. While some research support is available for aspects of grasp and bilateral skill development, very little research has been done with regard to object manipulation skills as a category within fine-motor development. Elliott and Connolly (1984) propose a detailed system for classification of manipulative skills, but the developmental sequence of these particular skills has not been studied. Ruff's work (1984) provides the most descriptive information about an infant's use of object manipulation skills. I have conducted some preliminary investigations of the development of in-hand manipulation skills with preschool and young school-aged children (Exner, 1990a, 1990b).

Development of a standardized test to assess children's in-hand manipulation skills will assist in the further development of knowledge about this performance area. The limited data collected in this study of the Test of In-Hand Manipulation by Exner support its further development. All aspects of construct validity considered in this pilot study yielded data that suggest the test is functioning appropriately. For example, the data show an increase in children's use of in-hand manipulation skills between the ages of 2 and 6 years and a tendency to be less well-developed in children with lower scores on a test of general fine-motor skills than in children with higher scores. The means for the test's scales appear to be reasonable in relationship to one another, based on my clinical observations of both normal and developmentally delayed children.

Continued development of the TIME will produce additional information about the development of specific in-hand manipulation skills. Eventually, the test will be useful for studying the relationship between manipulation and other areas of development such as attention, perceptual/cognitive skills, and daily living skills (including tool use) and children's play with small materials.

## Acknowledgment

Two of the studies by Exner (1990b, 1991) were completed in partial fulfillment of the requirements for a doctoral degree in Human Development from the University of Maryland. Support from the Towson State University's Faculty Research Committee was valuable in completion of two of the studies by Exner (1990a, 1991). Financial support from the Towson State University's Faculty Development Committee, a sabbatical from Towson State University, and a grant from the American Occupational Therapy Foundation supported the completion of the dissertation (Exner, 1991).

I thank those individuals who participated in the data collection phase of all of these studies. The Lida Lee Tall Learning Resources Center staff and the Towson State University Council Day Care staff were very cooperative in making arrangements for conducting these studies (Exner, 1990a, 1990b, 1991) with children in their programs. The 24 occupational therapists who served as raters in the content validity and reliability studies, and Beth Storch, an occupational therapy graduate assistant, are commended for their participation.

## References

Connolly, K. (1973). Factors influencing the learning of manual skills by young children. In R.A. Hinde & J. Stevenson-Hinde (Eds.), *Constraints on learning* (pp. 337-369). London: Academic Press.

Connolly, K., & Elliott, J. (1972). The evolution and ontogeny of hand function. In B. Jones (Ed.), *Ethological studies of child behaviour* (pp. 329-383). London: Cambridge University Press.

Corbetta, D., & Mounoud, P. (1990). Early development of grasping and manipulation. In C. Bard, M. Fleury, & L. Hay (Eds.), *Development of eye-hand coordination across the life span* (pp. 188-213). Columbia: University of South Carolina Press.

Elliott, J.M., & Connolly, K. (1984). A classification of manipulative hand movements. *Developmental Medicine and Child Neurology, 26,* 283-296.

Exner, C.E. (1989). Development of hand functions. In P.N. Pratt & A.S. Allen (Eds.), *Occupational therapy for children* (2nd ed., pp. 235-259). St. Louis: Mosby.

Exner, C.E. (1990a). In-hand manipulation skills in normal young children: A pilot study. *Occupational Therapy Practice, 1,* 63-72.

Exner, C.E. (1990b). The zone of proximal development in in-hand manipulation skills of nondysfunctional 3- and 4-year-old children. *American Journal of Occupational Therapy, 44,* 884-891.

Exner, C.E. (1991). Content validity, reliability, and generalizability studies of the test of in-hand manipulation by Exner. *Dissertation Abstracts International*. (University Microfilm No. ADG92-05057.9203

Gallagher, J.M., & Reid, D.K. (1981). *The learning theory of Piaget and Inhelder.* Austin, TX: Pro-Ed.

Gilfoyle, E.M., Grady, A.P., & Moore, J.C. (1981). *Children adapt.* Thorofare, NJ: Slack.

Holaday, B. (1985). Sensorimotor development in the presence of atypical object manipulation during infancy. *Maternal-Child Nursing Journal, 14,* 7.

Landsmeer, J.M.F. (1962). Power grip and precision handling. *Annals of Rheumatic Diseases, 21,* 164-169.

Levine, M.D., Carey, W.B., Crocker, A.C., & Gross, R.T. (1983). *Developmental behavioral pediatrics.* Philadelphia: Saunders.

Long, C., II, Conrad, P.W., Hall, E.A., & Furler, S.L. (1970). Intrinsic-extrinsic muscle control of the hand in power grip and precision handling. *Journal of Bone and Joint Surgery, 52-A,* 853-867.

Moss, S.C., & Hogg, J. (1981). Development of hand function in mentally handicapped and non-handicapped preschool children. In P.J. Mittler (Ed.), *Frontiers of knowledge in mental retardation* (pp. 35-44). Baltimore: University Park Press.

Napier, J.R. (1956). The prehensile movements of the human hand. *Journal of Bone and Joint Surgery, 38-B,* 902-913.

Parker, S.T., & Gibson, K.R. (1977). Object manipulation, tool use and sensorimotor intelligence as feeding adaptations in cebus monkeys and great apes. *Journal of Human Evolution, 6,* 623-641.

Ruff, H.A. (1984). Infants' manipulative exploration of objects: Effects of age and object characteristics. *Developmental Psychology, 20,* 9-20.

# 4

# Therapeutic Fine-Motor Activities for Preschoolers

*Carol Anne Myers*

## Introduction

The activities and suggestions included in this chapter were developed at the Brookline-Newton Early Childhood Collaborative in the metropolitan area of Boston, Massachusetts. This program serves preschoolers from 3 through 7 years of age with mild to moderate special needs. Services are provided to children who attend substantially separate, self-contained classrooms as well as to children who attend regular nursery schools. The children who are in regular nursery schools receive special services such as speech, occupational therapy, and physical therapy in their after-school hours. In general, the children who attend the program do not have severe neuromuscular conditions such as cerebral palsy, but tend to have learning disabilities and sensory integrative problems of a mild to moderate degree. The activities described in this chapter are used with all of these children, and could probably be modified for use with children who have more severe disabilities.

The theoretical rationale for the fine-motor program described in this chapter is based on the work of Mary Benbow, OTR, as gleaned from her workshops and book (see References). Her theoretical perspective has provided an invaluable foundation on which to base the work of the program. Many of her ideas for fine-motor activities with older children have been modified and incorporated into the work with preschool children.

Most of the children in the program who receive occupational therapy services receive them once weekly, for 30 to 45 minutes, in either an individual or small-group setting. In addition, the occupational therapist consults with the parents and the classroom teacher. Surrounding the child with a team of people who are familiar with the child's strengths and weaknesses and with the goals of the therapeutic program greatly enhances the therapy process.

## Specific Activities for Fine-Motor Development

### Vertical Surfaces

The significant role of vertical work surfaces cannot be overemphasized when discussing the development of appropriate hand and wrist position for fine-motor and handwriting skills. Mary Benbow (1990d) emphasizes the importance of working on a vertical surface. In this posture, the wrist is correctly positioned to develop stability and support thumb abduction and opposition for developing dexterity. Stable wrist extension and thumb opposition also facilitate total arching of the hand for skillful manipulation of objects.

Finally, wrist extension seems to facilitate balanced use of the hand's intrinsic muscles (Benbow, (1990b). This wrist alignment enables the thumb to move into its opposing plane to work distally with the finger pads (Kapandji, 1982). Also, this wrist alignment facilitates balanced use of the hand's intrinsic (fine-motor) muscles (Benbow, 1990b). Providing a vertical work surface is, therefore, an important modification that parents and teachers can incorporate as they work or play with the child.

Not only does the vertical surface promote wrist extension, but also the development of arm and shoulder muscles as well. Rather than having preschoolers work while leaning over small tables, parents and teachers are encouraged to provide activity areas in which the children are working upright (either sitting or standing) with their arms and hands moving against gravity at an easel or other vertical work surface. When children work on a horizontal surface, they often place their wrists in neutral or in flexion, which are not positions that promote skillful use of the intrinsic muscles. Switching activities from a horizontal to a vertical orientation can transform an ordinary or mediocre activity into a powerful tool for encouraging fine-motor skill development.

Many activities can be placed on the vertical by using book holders on a table, tabletop easels, or regular floor easels. In the program, children are expected to work daily on the easel, and therapists encourage teachers to ensure that any activity that can be performed on the easel be routinely done that way. With a minimal amount of modification and equipment expense, a tremendous number of activities can easily be adapted for use on a vertical surface. Following are activities that have been found helpful.

1. Making pictures with stickers (Stephens' "Play Shapes" stickers from an educational toy store are particularly motivating, as are the large sheets of circular stickers available in bulk from office supplies stores. Other kinds of stickers that appeal to children may also be used).
2. Colorforms or Unisets (these activities provide a board on which to arrange reusable plastic "stickers," and they are available in a wide variety of themes and designs).
3. Feltboards or flannel boards, which permit the placement of figures depicted in stories or scenes created by the child.
4. Chalkboards (sidewalk chalk, which is wide-diameter chalk, should be broken into 1 1/2- to 2-inch pieces for children to hold with the tips of the thumb, index finger, and middle finger. In one favorite activity, the child draws a design with the chalk, and then uses a paint brush with water to "magically" erase the design).
5. Geoboards (rubberband designs on a grid of nails).
6. Puzzles with thick pieces having small handles.
7. Magna-Doodle (this is a "magic" writing board that uses a magnet pencil and magnet shapes to form the marks on the board. When using this toy on a vertical surface, the board should be turned upside down so the "erase" lever is at the top instead of the bottom. Children can use the magnet pencil to draw freehand, and a Stetro pencil grip can be fitted onto the pencil if necessary. The magnet shapes can be used to help promote a pad-to-pad grasp of the thumb, index finger, and middle finger. Magna-Doodle accessory kits with additional magnet stamps and tracing cards can be purchased for this toy to make it more interesting and challenging. For example, there is a zoo accessory kit with zoo scenes to trace and animal magnet stamps to then stamp onto the traced picture. Other theme kits such as circus and Mickey Mouse are also available, as well as a spirograph kit).
8. Painting or drawing.
9. Ink-stamping activities.
10. Pegboards, many different varieties, including Lite-Brite.
11. Magnet letters or shapes on a magnet board (some educational companies feature delightful wooden figures on magnets that are familiar characters from stories such as *The Three Bears*. These story figures are used on a vertical metal surface and are incorporated into the speech and language group activities. Magnatooli, by Anatex Enterprises, includes a metal board with colorful magnet shapes for forming designs and pictures).

If there is empty wall space in a home or classroom, a large piece of felt can be attached to the wall, and children can use felt shapes or flannel board accessories to make pictures. One teacher has each child sign in on a clipboard attached to an easel every morning on arriving at school. The preschool teachers also have "circle time" boards (with the weather and calendar) placed on vertical surfaces. This surface can be used at a different time of the day for another activity. Preschool teachers often cover wall-mounted chalkboards with bulletin board decorations or children's projects, but the lower portion of the chalkboards can easily be cleared for children's use. Some educational equipment catalogues now feature low storage shelves that have double easels installed on the top, at just the correct height for preschool children. Lite Brite is a game produced commercially that is on the vertical, and one teacher uses the Lite Brite stand for holding other kinds of activities that are appropriate for a vertical surface. Directions for making a cardboard easel are included (see Appendix). One nursery school with a small budget made several of these easels and distributed them throughout the classroom. The Appendix also includes information about ordering the tabletop easels that are recommended to parents and teachers, as well as information about how to obtain some of the items listed above.

The Collaborative therapists encourage the preschool teachers to use easels in their rooms, and encourage the parents of children seen by occupational therapy to provide them at home. As part of the demonstration to parents or teachers, children are often asked to perform an activity such as a pegboard while on a horizontal surface, and then to perform the same activity on a vertical surface. The difference in the child's hand position and ability is often dramatically evident in such a demonstration, and it helps parents and teachers understand why working on the vertical is so valuable. Throughout this chapter, the examples will emphasize how a variety of therapeutic goals can be implemented through activities placed so the child is working on the vertical, thereby maximizing the therapeutic benefit of any particular activity.

## Manipulatives

Young children, especially 3- and 4-year-olds, should spend more time with fine-motor *manipulatives* than with writing utensils. Sometimes parents and teachers feel that young children should begin to "practice" with pencils and markers, but this early practice may result in a poor pencil grasp, partially because children may be asked to use writing utensils before their hands are ready for that kind of refined activity. Children should be developing their hands for a *variety* of activities in a *variety* of positions before they are expected to draw or write with the proper grasp.

Pencil postures fixed early by repeated use at an intermediate level of skill may have a negative impact on graphomotor performance when speed and volume demands increase (Benbow, 1990b). In preparation for writing, the hand progresses through the following motor milestones (Benbow, 1990d).

1. Development of palmar arches in the hand; this means that the child will develop a concave surface on the palm generally indicating that the palmar arches are present.
2. Development of wrist extension to support skilled finger movements.
3. Development of an awareness of the "skill side" of the hand; this means that the child will consistently orient skill activities toward the thumb, index finger, and middle finger. These three fingers will hereafter be referred to as the "skill fingers."
4. Development of an open index finger-thumb web space when performing skilled activities. The open web space should have a circular shape. This position is frequently compromised in children who have hyperextension of the interphalangeal joint of the thumb; rather than a circular web space, these children form a "crescent moon" with a small opening. The thumb is in a fixed position, thereby making intrinsic muscle activity difficult (see Figure 1). Children with this problem must be carefully watched when they perform fine-motor activities in order to find those activities that encourage the use of the thumb in a flexed position.
5. Development of intrinsic muscle movement in the fingers; this kind of fine-muscle movement can be seen when the ulnar side of the hand is stabilized on the table while the fingers move a pencil to write, or when the fingers make fine movements to thread a needle. The intrinsic movements are best observed in activities that require the tips of the thumb, index finger, and middle finger to be touching while they are performing small movements of mid-range flexion and extension of the metacarpal-phalangeal (MCP) joints. Many so-called "fine-motor" activities involve the use of the hands and fingers, but do not necessarily elicit the fine-motor movements of the intrinsic muscles at the MCP joints.

**Figure 1. Hyperextended Thumb on Lace Tip**

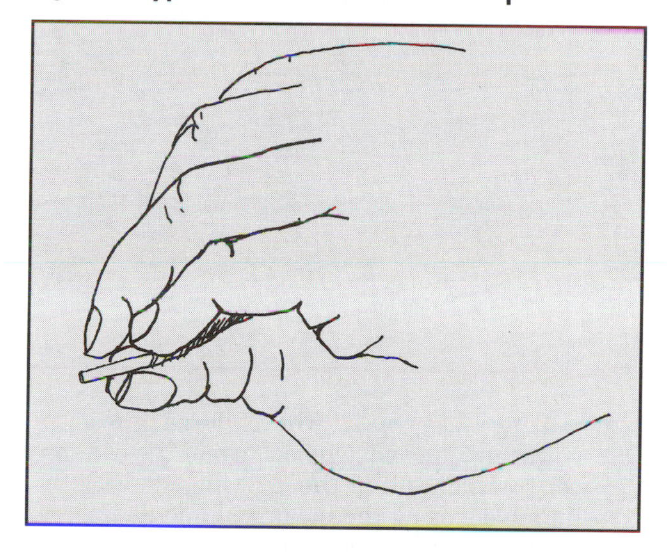

Below is a list of sample manipulative activities that are used in occupational therapy sessions with the children. When appropriate, these activities are also suggested to parents and classroom teachers. The most successful outcome of a treatment session is for children to be so interested in an activity that they request to repeat or continue the activity at home! The area of hand development that the activity is most likely to encourage is noted in brackets after each activity.

1. *Playdough:* Playdough is a wonderfully inviting activity that most children enjoy. Children are encouraged to mold and roll playdough into balls with the following motion: The playdough is manipulated between their palms, with their fingers curled slightly toward the palm, and with flexion of the carpal-metacarpal joints [ARCHES]. Children can also roll playdough into tiny balls ("peas" or "marbles") by using only the fingertips of the skill fingers in a pad-to-pad orientation [SKILL SIDE, INTRINSICS]. Playdough can be flattened on the table or on a vertical easel surface, and the child can use a

small peg or toothpick to make designs in the playdough [SKILL SIDE, WEB SPACE, INTRINSICS]. When cutting playdough, the child should be encouraged to hold the plastic knife or pizza wheel with a diagonal volar grasp [ARCHES] (see Figure 2). (See Appendix for information about a book that provides interesting ideas for playdough activities.)

4. *Tweezers and scissors tweezers:* Oversized tweezers such as those found in the "Bedbugs" game by Milton Bradley Company can be held with the thumb opposing the other fingers, or in a tripod grasp. The children can use the tweezers to pick up small objects such as Cheerios or small cubes as part of a counting activity. Another alternative is to use scissors tweezers with the correct scissors grasp to pick up cotton balls or tiny toys. One teacher created a popular game called "Tiny Dinosaur Pick Up" in which several children were given a pair of scissors tweezers and they raced to see how many dinosaurs they could pick up and put into a tray in a given amount of time [ARCHES, WEB SPACE, INTRINSICS].

**Figure 2. Diagonal Volar Grasp**

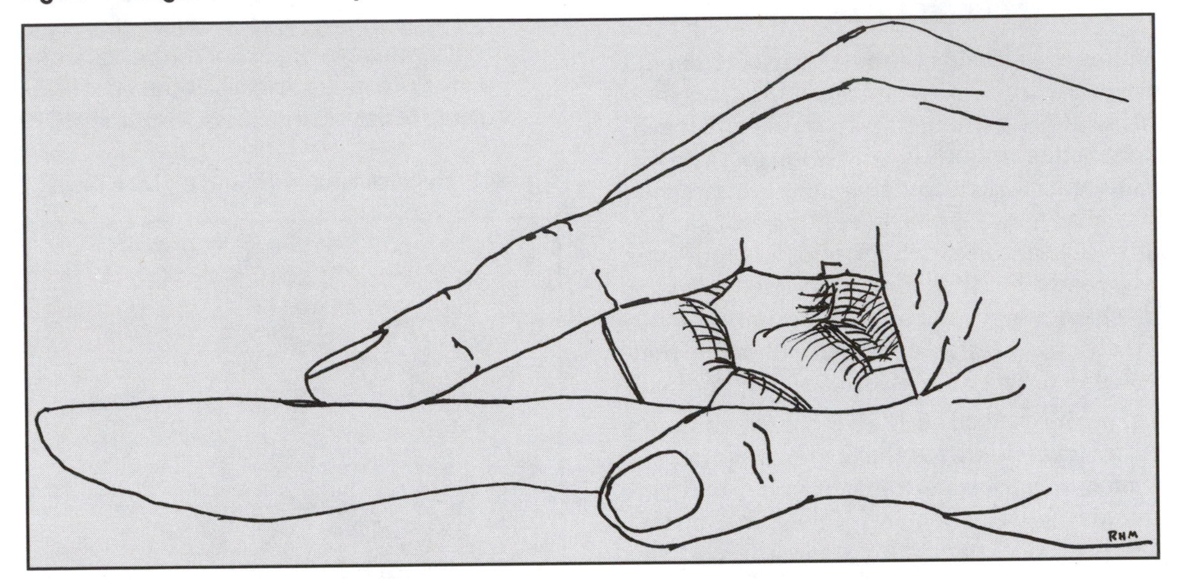


2. *Tearing newspaper:* The children are given newspaper and shown how to tear the newspaper by holding it in the skill fingers, with the thumb opposing the index and middle fingers. After tearing the newspaper into strips, they crumple it into balls and use it to "stuff" a scarecrow, snowman, or other art creation. An entire classroom of children can participate in decorating two large dinosaur pictures that can then be taped or stapled together back to back, and stuffed with crumpled newspapers to create a marvelous three-dimensional creature to hang from the ceiling of the classroom [ARCHES, SKILL SIDE].
3. *Plant sprayers:* Plant sprayers are frequently used for a wide variety of activities. Sprayers with large handles open the web space and encourage the use of the fingers in an opposed grasp. Using water guns with small handles can encourage development of the skill side of the hand, since only the skill fingers will fit onto the handle. Children use plant sprayers to spray plants (indoors and outdoors), to spray designs onto snow outside in the yard or playground (mix food coloring with the water so the snow can be "painted"), or to "melt" monsters (draw monster pictures on an easel with markers and the colors will "melt" when sprayed) [ARCHES, SKILL SIDE].
5. *Dice games:* Many young children have difficulty cupping their hands in order to shake dice. Often this appears to be due to poor palmar arch development. Children can be taught to cup their hands to form a little "house" for the dice, and are gratified when shaking the dice produces a rattling sound of success. Teachers can easily incorporate dice games into classroom activities for counting tasks [ARCHES].
6. *Lacing activities:* Lacing activities can provide an excellent challenge for children developing hand skills. Laces with small tips require the most skill, as do lacing cards with small holes. Small tips on the laces encourage the child to use just the index finger and thumb in a pad-to-pad grasp on the tip, and the small holes in the lacing cards encourage intrinsic muscle movement of the fingers. Children can also string cut-up straws, beads, Cheerios, and other small objects.

When stringing beads, large beads are sometimes more difficult than smaller ones since the child must use intrinsic muscle movements to work the string all the way through the bead hole. Although small beads are more difficult to hold in the fingers, they are easier to lace since one thrust of the lace will bring it all the way through the hole. When task analyzing a lacing or bead-stringing activity, one must understand the goals of the activity: to teach the general concept of bead stringing, develop bilateral use of the hands, or improve intrinsic muscle movement. The choice of beads or lacing materials might vary considerably, depending on what skill is to be taught.

Another activity that is similar to lacing is having the child insert small feathers into oaktag (light-weight poster board available at art supply stores) cut out in the shape of a mask. The child decorates the mask, which has slits cut into it for the feathers. While holding the tip of a feather in the skill fingers, the child inserts it into a slit [SKILL SIDE, INTRINSICS] .

7. *Eye droppers:* Eye droppers can be used at a water table in the classroom or in the bathtub at home. Water can be mixed with food coloring to make "dribble pictures" by dripping the food coloring onto paper towels [SKILL SIDE, WEB SPACE].
8. *Tissue paper pictures:* The therapist gives the children scraps of tissue paper, and asks them to roll each piece into a small ball by using only the skill fingers. The balls can be glued onto construction paper to form a picture. Sometimes the therapist can draw a general shape (e.g., a pumpkin outline) and the children can make enough tissue paper balls to fill up the outline [SKILL SIDE, INTRINSIC MOVEMENTS].
9. *Coins and buttons:* Children can play a variety of games with buttons and coins, including using the skill fingers to insert them into a bank, picking them up and arranging them as part of a counting or matching game, making designs with buttons on the table, ordering buttons according to size, and so on. Teachers in the Collaborative often use button activities to reinforce academic concepts while challenging fine-motor skills. The buttons or coins must be moved or turned over *without* bringing them to the edge of the table. The teachers and therapists have large containers of mixed buttons obtained from a local fabric store that provide materials for numerous creative and inviting activities for both the treatment sessions and the classroom [SKILL SIDE, INTRINSICS].
10. *Cooking activities:* When making cookies in the classroom, the teachers encourage hands-on participation for the children. When putting sprinkles onto cookies, the sprinkles are poured into a bowl so that the children have to pick them up with their fingertips. The children also participate in tearing lettuce, pressing out pizza dough, and other kinds of food preparation using their fingers [SKILL SIDE, INTRINSICS].
11. *Games with the "puppet fingers"* (the skill fingers): At circle time, the teachers ask each child to use his or her puppet fingers to tell about what happened over the weekend or to describe their "Show and Tell." Many of the teachers have incorporated puppet fingers into speech and language activities. Children are encouraged to keep their ring and little fingers motionless and curled into the palm while they move the skill fingers independently, which encourages separation of the two hand sides for skill. The finger position and motions during the puppet fingers activity are similar to the finger position and motions employed while using scissors [SKILL SIDE].
12. *Tomy Waterfuls games:* The Waterfuls games are made in two attached sections: an upper section filled with water and small objects that comprise the moving parts to the game, and a lower section with a large button. Pushing the button causes motion in the water, which moves the objects through the water tank. Careful timing in pushing the button causes the objects to land in the correct spot to win the game. For example, in the Ring Toss game, a favorite in the occupational therapy clinic, rings move through the water as the player attempts to make them land on poles.

The Waterfuls games encourage children to place their hands in an open web space posture with their thumb in opposition to the other fingers. A child must apply pressure with the thumb while it is in opposition, and maintain this hand position throughout the game. When playing this game, it is important to watch for hyperextension of the thumb interphalangeal joint and to encourage the child to flex this joint when pushing the button [ARCHES, WEB SPACE, OPPOSITION GRASP].

Although all of these activities encourage the development of the muscles needed for fine-motor skill, the therapist needs to attend to *how* a child performs them. A child with poor hand skill will always find a way to use the less-skilled lateral pinch grasp, even in the best-designed activity (see Figure 3).

Once the skill component to be developed is identified, the therapist can then refer to the activities list to determine those activities that are likely to be most successful with the child. Since it is somewhat difficult to imagine how a child will perform an activity, given only a written description as above, creative experimentation with the above activities will help determine their appropriateness for different kinds of clients. The above list is certainly not a comprehensive list of all the possible activities, but it includes the favorites and the ones most likely to be successful with many children. Also, the above activities are meant to be more than a list of "best" activities; they are also intended to be "seeds" that will help guide the ongoing selection of a wide variety of therapeutic activities and toys. New activities ideas are constantly being contributed by the teachers, parents, and children, and many of the traditional nursery school activities (e.g., gluing pasta and beans to make pictures) will provide the same kinds of challenges as those activities presented above. When working with teachers, therapists not only suggest new activities, but point out those activities and toys that they *already* use that help children to develop good hand skills.

## Scissors

When scissors are held correctly, and when they fit a child's hands well, cutting activities exercise the same intrinsic muscles that are needed to manipulate a pencil in a mature tripod grasp. When scissors are held incorrectly, cutting activities are performed primarily by the larger muscles of the forearm rather than primarily by the intrinsics (Benbow, 1990c). The correct scissors position is with the thumb and *middle* finger in the handles of the scissors, the index finger on the *outside* of the handle to stabilize, and fingers four and five curled into the palm. The lower handle of the scissors should rest on the distal interphalangeal joint of the middle finger, and the upper handle of the scissors should rest on the distal interphalangeal joint of the thumb. The tips of the scissors should be pointing away from the child, and the wrist of the cutting hand should be in extension (Benbow, 1990d). When cutting, movements of the fingers should be in the intermediate range of excursion between very flexed and very extended in order to use the intrinsic muscles to their maximum benefit (Benbow, 1990c) (see Figure 4).

Many children hold scissors with the thumb and index finger in the handles. This position does not allow for proper control of the scissors, and does not help develop the hand for fine-motor skill. Parents and teachers can make a tremendous difference in a child's hand development simply by teaching the proper scissors grasp. Therapists need to monitor children throughout the year to evaluate whether or not they continue to use the correct grasp since in the early

**Figure 3. Lateral Pinch Grasp**

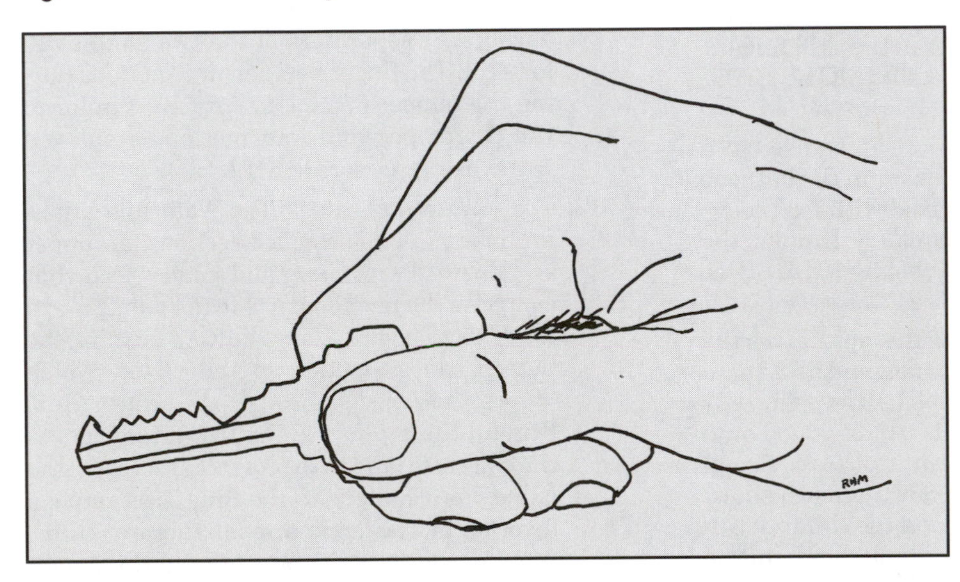

**Figure 4. Scissors Grasp**

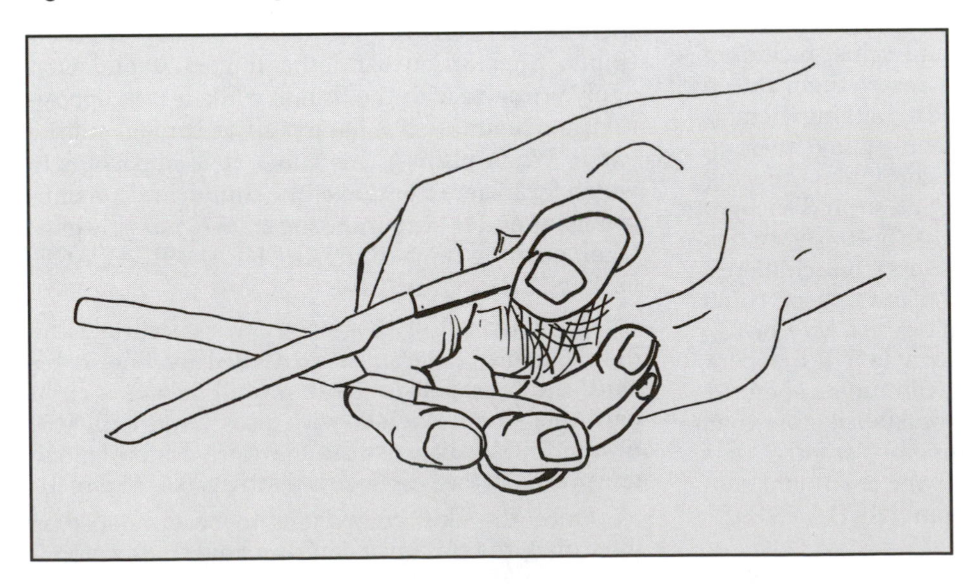

stages of learning the habit can be lost. The best scissors for children are sharp, and have small-holed handles. (See Appendix for ordering information on children's scissors.)

Cutting with scissors can be an excellent fine-motor activity, and scissors activities can be adapted to children of varying skill levels. Four years of age is the appropriate time for the majority of children to begin learning scissors skills. The hands of most 3-year-old children are so small that even the tiniest scissors available have handle holes that are too large to allow for proper control with the correct grasp. When the handle holes are too large, children tend to place most or all of their fingers into the handles, thereby learning the incorrect hand position for skilled use of scissors. Three suggestions for teaching beginning scissors skills follow; for additional suggestions, see Schneck and Battaglia, this volume.

1. Children who are first learning to use scissors should cut card weight junk mail, because it is an activity that is easily successful. Many children are thrilled to make a huge pile of tiny pieces of paper. For beginners, just holding the correct cutting position of the hand and making open and close movements are a challenge. Cutting junk mail provides a "product," and one with a 100% chance of being acceptable. Scissors cutting at this level of development is analogous to the "scribbling" stage of learning to draw. Another activity at this level of development might be to have the children "fringe" the edge of a piece of paper. Again, only open and close movements of the scissors are required.
2. Children can cut leftover scraps of construction paper into random pieces, and can then use a glue stick to glue the pieces on another paper. With this activity, children often begin to experiment with trying to cut out specific shapes, but without the challenge of having to cut on the lines. Scissors cutting at this level of development is analogous to early representational drawing activities, when a random movement results in a meaningful outcome. With cutting as with drawing, children begin to learn the relationship between finger movements and the results they produce. To make the overall activity more therapeutic, the paper can be placed on a vertical surface while gluing, and the glue stick can be held in a tripod grasp, just like a pencil or marker.
3. The instruction book, *Learn to Cut* (Wolfe, 1987), is an excellent resource for early cutting activities. It is developmentally sequenced, with a wide range of graded cutting activities that are intended for reproduction on a copying machine (see Appendix for ordering information).

## Drawing and Writing

When using drawing or writing implements, children are expected to use either a tripod or quadripod grasp. A number of researchers describe components of an effective pencil grasp, as summarized by Benbow (1990b). The most important component of an effective pencil grip is a rounded open web space between the thumb and index finger (Long, Conrad, Hall, & Furler, 1970). Initially, the thumb and fingers hold the pencil while the wrist and elbow provide the movement ("static tripod" grasp). Later, the fingers and thumb flex and extend in alternation, resulting in distal dynamic control (Rosenbloom & Horton, 1971). A standard three-digit pencil grip, the "dynamic tripod," allows the fastest, longest, and most dexterous control of a pencil by the human hand (Connolly, 1973). Approximately one half of all normal children follow an alternate four-finger "quadripod" progression (see Figure 5) (Benbow, 1987; Benbow, 1990b).

The normal sequence of development, therefore, seems to be that children initially use a static grasp (tripod or quadripod), and progress to using a dynamic grasp (tripod or quadripod). The majority of nondysfunctional children (57.5% by 3.5 years and 77.5% by 6.11 years) use either a static tripod or quadripod grasp, or a dynamic tripod grasp (Schneck & Henderson, 1990). A much smaller percentage

**Figure 5. Tripod Grasp**

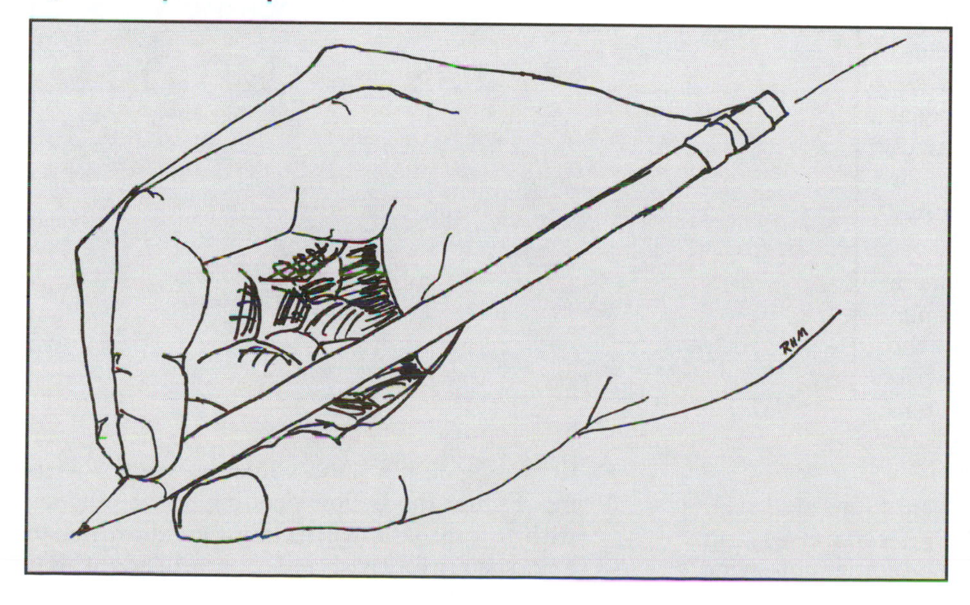

of children use the "lateral tripod grasp" (7.5% by 3.5 years and 22.5% by 6.11 years) (Schneck & Henderson). Further research should be done in order to establish whether or not the lateral tripod grasp is a desirable grasp, particularly since it does not include the open web space component of the static and dynamic tripod grasps (see Figure 6).

Figure 6. Lateral Tripod Grasp

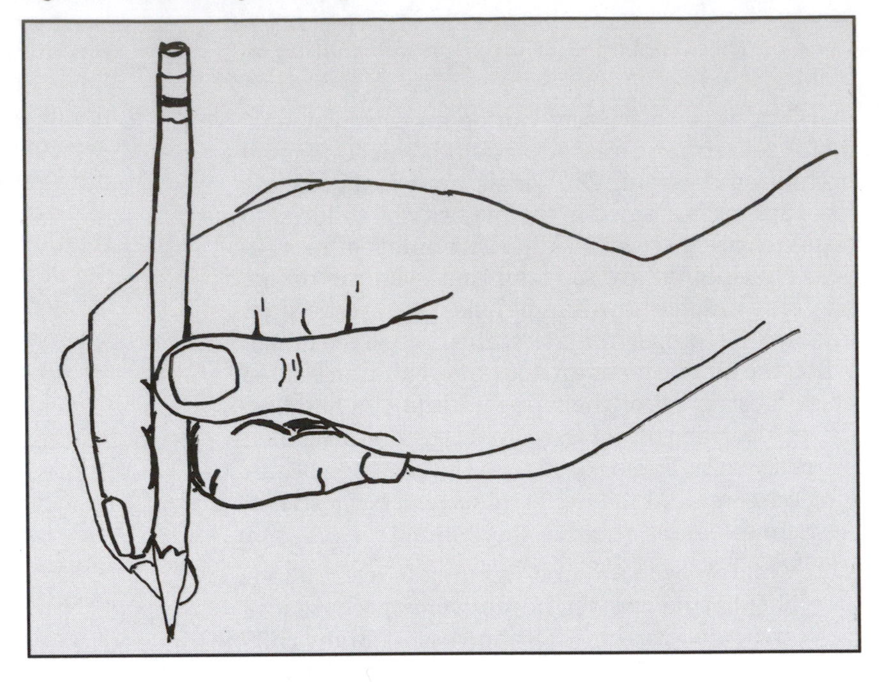

Since preschoolers are at a malleable stage of fine-motor development, and because the preschoolers referred to the Collaborative are considered to be "at risk," the Collaborative therapists and teachers encourage children to use either a tripod or quadripod grasp. These open web space grasps are used to perform common activities of daily living, such as buttoning small buttons. Individual variations in pencil grasp may occur as the children continue through later grades in school, but hopefully those variations will contain the important components of the dynamic tripod grasp, as described above by Mary Benbow. However, some children do not ever achieve an "ideal" grasp; for example, one child with a mild neuromuscular disorder was not able to sustain an open web space posture even after years of intervention. In cases such as these, the therapist must be able to help the child develop his or her optimal skill level with an alternative grasp to the best of his or her ability.

Figure 7. Grasp on Preschool Crayon

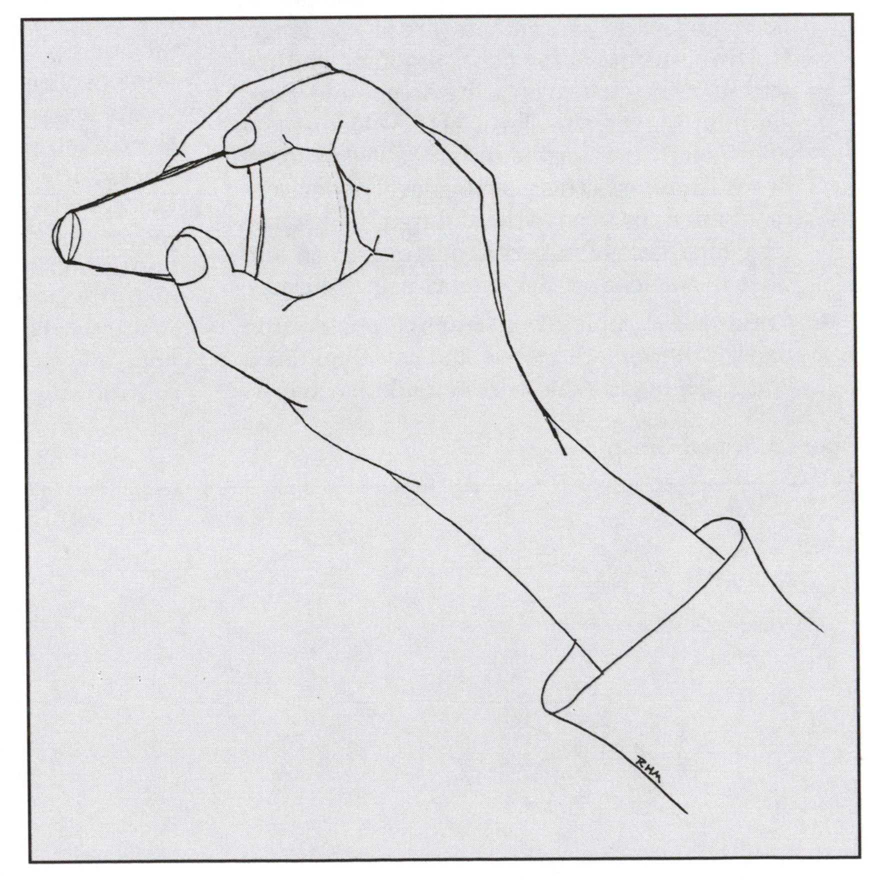

Collaborative children who have had intervention for fine-motor delays tend to develop the dynamic component to their pencil grasp later than normal children. The therapeutic goal for these children is that they have a solidly developed static tripod or quadripod grasp by the time they enter kindergarten.

Therapists may choose from a variety of drawing and writing implements when helping a child make the transition from manipulatives to writing utensils. The following list of utensils and activities is hierarchically arranged by developmental levels.

1. *Preschool crayons:* When held in an open web space posture and used on a vertical writing surface, these crayons provide static support for the young child's hand while drawing or writing. Filling the hollow opening of the crayons with cotton balls will remind children not to hold the crayon with one finger inside it, but

rather, to keep their fingers on the outside surface. (Drawing with one finger inside the crayon tends to put the thumb-index finger web space in a closed position.) The only drawback to these crayons is that they provide a great deal of resistance in order to make a mark on the paper, which is frustrating to many children (see Figure 7). (See Appendix for specific product information.)

2. *Primary-sized markers:* These are one of the most effective writing implements for children who are learning to use the correct finger position for drawing or writing. Children use these markers while drawing on easels, and are encouraged to use their "tips [of their fingers] on the stripe" at the base of the marker. Due to the light resistance of the markers when moved across paper, children with weak hands can sustain a static tripod grasp easier with primary-sized markers than they can with crayons or primary-sized pencils. (See Appendix for specific product information.)

3. *Stencils:* Many of the children in the program are reluctant to attempt representative drawing, but are willing to use stencils for their first pictures. They use basic shape stencils such as circles, squares, and triangles to draw parts of robots and monsters, which are "right" no matter how children draw them. As the children gain confidence, they are encouraged to draw robots and monsters without using the stencils. (When markers are used, children can "melt the monster" using their spray bottle! (See "Manipulatives" activities, above.)

4. *Duo drawing:* Many children enjoy drawing as a partner with the therapist. This means that the therapist and the child take turns completing one drawing. The therapist might draw the outline of a snowman and make one of the eyes. The child then draws another eye, the therapist draws the nose, and the child draws the mouth. When the therapist provides the initial outline, and then takes turns with the child, children who are reluctant to attempt drawing are more willing to participate in this activity. This is often an activity that parents can do at home with their children.

5. *S.O.S.:* This version of S.O.S. is similar to the original version, except that initials are not used in the squares. The child and therapist each choose a differently colored marker, and one person starts the game by drawing a vertical or horizontal line between two adjacent dots. The next person draws a line between two dots, and the players keep taking turns drawing lines in an attempt to finish a square. The person who draws the fourth side of any square is allowed to make a dot inside that square, thereby marking it as his or hers. Once all the squares in a grid are completed, each person counts his or her dots and a winner is declared. This is an excellent prewriting game for teaching pencil control, starting and stopping ability (needed for printing letters), and for encouraging top-to-bottom and left-to-right formation of writing strokes. It can also encourage top-to-bottom and left-to-right sequencing when the therapist or teacher helps the child organize his or her counting of the dots to determine the winner. Children can develop some strategy skills as they begin to learn how to plan their move so that their opponent's next move will not finish a square. S.O.S. grids can vary widely in size, but a 16-dot grid seems best for most preschoolers (see Figure 8).

6. *Tracing:* The act of carefully tracing the outline of a drawing often elicits a more skilled grasp than the act of coloring the drawing. Children are asked to perform tracing activities as another therapeutic activity to enhance the development of prewriting skills.

**Figure 8. (left) Blank S.O.S. Grid; (right) Grid: Game in Progress**

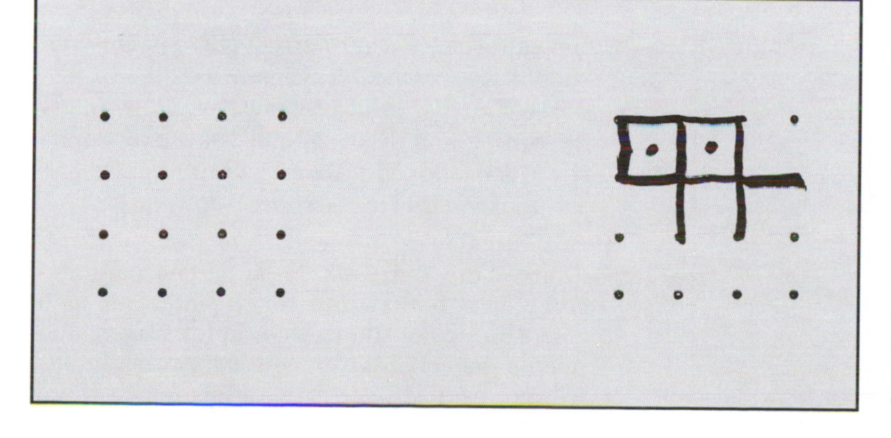

When children are ready to begin learning the proper grasp on a writing implement, they begin by using primary-sized markers on a vertical surface. The therapeutic activities are very short in length at first, because the child is expected to sustain the proper grasp for the entire activity. The therapists use specific verbal and visual cues to help the child remember to keep the tips of the thumb, index finger, and middle finger on the stripe close to the tip of the marker, and to keep the marker well-positioned into the thumb-index finger web space. Some children respond well to the use of the word "tips," and this cue word has successfully been transferred from occupational therapy treatment sessions to use by the teachers in the classroom. One teacher used the visual cue of puppet fingers ( formed with the teacher's own fingers) to signal a child to reposition the marker correctly. This was particularly effective for a child who was embarrassed by verbal reminders, but who was comfortable with the visual cue. Another child who was adept at using darts learned the correct grasp by reminding himself to hold the marker the same way he held darts. Another child found that the easiest strategy while learning the correct grasp was to verbalize "The marker rests in the web space," as she positioned the marker in her fingers. As children gradually become more comfortable with an appropriate static grasp, the amount of time they are expected to use that grasp during an activity is lengthened. Parents and teachers are encouraged to begin to ask children to use the grasp they learned in occupational therapy drawing activities at home and at school.

The problem of thumb interphalangeal hyperextension, which was mentioned in the 'Manipulatives" section, can also create problems for the child who is trying to learn the correct pencil grasp. When the thumb is hyperextended, it "fixes" the half-closed web space position so that intrinsic muscle movement is difficult to achieve (see Figure 9). Children should be taught to keep their thumb in flexion on all writing implements to facilitate a fully opened web space posture.

Some children who are chronologically ready to enter kindergarten do not have the hand skills expected for a child their age. In an ideal world, a child with poor hand skill would not be expected to use writing implements until his or her hands were truly ready for skilled finger tasks, but sometimes it is necessary to superimpose the learning of a static tripod or quadripod pencil grasp onto an unskilled hand. Children with poor hand skills can sometimes develop an appropriate static tripod or quadripod grasp, but continue to need to have their array of daily activities enriched by the use of therapeutic manipulatives since their use of the static grasp (as opposed to a dynamic grasp) does not enhance the formation of small-sized printed letters nor does it enhance the child's ability to write in cursive.

**Figure 9. Pencil Grasp Showing Thumb Interphalangeal Hyperextension and Compromised Thumb–Index Finger Web Space**

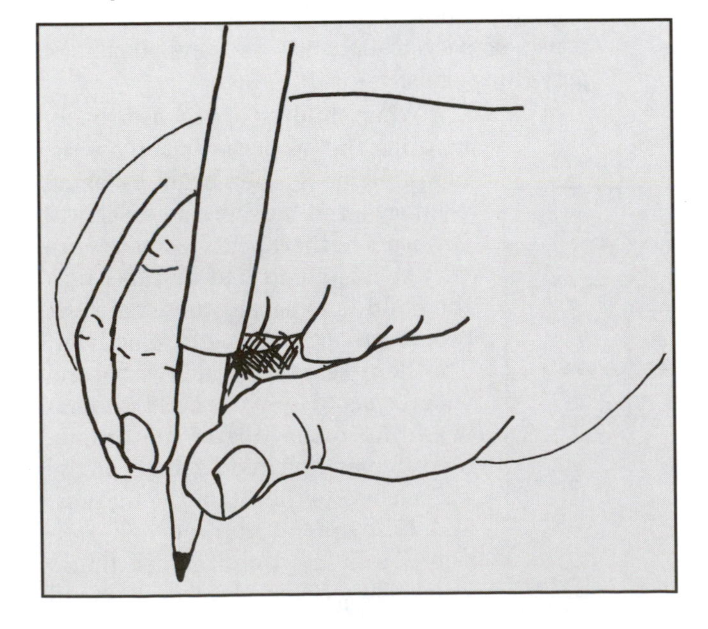

## Activities By Ages

The clinical experience of the Collaborative therapists indicates that some children who have an immature or incorrect pencil grasp may have used pencils or crayons at an early age when, for those particular children, their hands were not ready for that kind of fine-motor task. Fine-motor skills develop at different ages in different children, and the following guidelines are intended to present a conservative approach to providing drawing and writing implements for children. In this way, children whose hands are not yet ready for pencils and crayons can participate in appropriate manipulative activities, thereby preventing their developing an incorrect grasp in their attempts to use an implement that is too sophisticated for their level of development. Although it is preferable to have children use a variety of manipulatives to encourage overall fine-motor development, most children have a strong need for an outlet to express themselves through scribbling and drawing. This list was designed to provide suggestions to encourage skill development by using fine-motor materials that do not interfere with the child's existing skill level. The suggested activities should be incorporated in therapy or classroom activities several times a week.

### 3-Year-Olds

1. The 3-year-old should use a vertical surface for activities on a daily basis.
2. Instead of pencils, markers, or crayons, children should use preschool crayons on a vertical drawing surface. For children whose hands are ready for writing and drawing, and who have a solid tripod or quadripod grasp on a writing implement, the primary-sized markers are a good choice.
3. In addition to the preschool crayons, some children can use "sidewalk chalk" (oversized-diameter chalk), broken into 2-inch pieces and held with the tips of the thumb, index finger, and middle finger. Children with very weak hands will not be able to sustain an open web space

grasp with chalk; rather, their fingers will collapse into an undesirable fisted clench with a closed web space. Therefore, the child's hand position should be monitored to determine whether or not chalk is an appropriate activity for him or her.

4. For those children who wish to make pictures, gluing small pieces of construction paper or using stickers are alternatives to writing utensils.
5. Painting on easels is also recommended, if the child holds the brush in a tripod or quadripod grasp.

## 4-Year-Olds

1. All of the suggestions for 3-year-olds are appropriate to continue with 4-year-olds.
2. Scissors activities should begin, and children should be taught the correct scissors grasp from the first moment they use scissors (thumb and middle finger in the handles, index finger on the outside to stabilize).
3. Crayons are appropriate for children with well-developed hands, but children with weak hands usually perform better with primary-sized markers. Frequent monitoring of the child's hand position will indicate which drawing and writing implements are most appropriate for the child's developmental skill.

## 5-Year-Olds

1. Most 5-year-olds use either a static or dynamic tripod or quadripod grasp (Schneck & Henderson, 1990). By the end of kindergarten, most children can write or draw using intrinsic muscle movements of the fingers with their hands stabilized on their papers, rather than drawing with wrist or arm movements. It is appropriate for these children to use crayons, pencils, and markers for drawing and writing activities. Five-year-old children can benefit from working on vertical writing surfaces when possible, although many children have well-established grasps with adequate wrist extension at this point, and can successfully sustain their grasps when working on horizontal surfaces.
2. For children who do not use the proper pencil grasp, suggestions from the previous two levels of development should be incorporated into their daily activities. If a child's hand has not progressed to be ready for pencils and crayons, they should continue to use earlier kinds of writing implements to prevent their developing a dysfunctional grasp. For example, allowing a child to use pencils or crayons with a fisted grasp does not promote use of the correct grasp, nor does it facilitate development of the appropriate musculature to support it.
3. When children must use pencils for the required school work, yet their hands are not quite ready to sustain a mature grasp, sometimes the use of a pencil grip such as the "Stetro" pencil grip can be helpful (see Appendix for ordering information). To encourage consistent use of a writing implement fitted with the Stetro, parents can purchase either a set of Crayola color pencils or Sanford's plastic markers, and also purchase enough Stetros to fit onto all the colors in the set. The child is consistently encouraged to use this set of colors (and also a regular pencil fitted with a Stetro) until he or she has solidly developed the correct grasp.

One occupational therapist discovered that some kinds of bulletin board pushpins will fit into the Stetro grip, so the child can hold the pushpin using the Stetro and make pictures by pressing the tip of the pushpin into construction paper layered over cardboard (and placed on a vertical surface if possible). With the ulnar side of the hand stabilized on the paper, the small finger movements needed to prick the holes in the paper are the same intrinsic movements needed for writing with a pencil.

Consultation with an occupational therapist will help teachers to determine the stage of development of the child's hands, and the occupational therapist can guide the selection of activities for the child at school and at home. There are a variety of "transition grasps" that are quite appropriate, but that might appear awkward to the untrained parent or teacher. The occupational therapist can determine whether a child's grasp is in transition toward a mature grasp pattern, or representative of an immature stage of development that requires the use of writing implements from earlier age levels.

One effective tool for enhancing the understanding of written evaluations and reports is to attach a line drawing of the hand position being described. Occupational therapy reports frequently use terms that parents and teachers do not understand, and sometimes hand positions are difficult to accurately describe. Drawings such as the ones included in this chapter are photocopied, cut out, and taped within the body of reports to help educate parents and teachers about both functional and dysfunctional hand positions as they relate to an individual child. Through the use of selected drawings, each report becomes an individualized account of a child's performance level and expectations for future development. The drawings included in this chapter can be photocopied by the reader, and included in reports.

## What Makes Therapy Effective?

Clinicians at the Collaborative do not have the absolute answer to this question, but have intuitive insights based on their years of working with children with fine-motor problems. Three-year-old children are not usually aware that they have a fine-motor problem, so the treatment planning approach is different for them than for older children. For example, the 3-year-old needs activities that are so intrinsically motivating that the child is not cognizant of how challenging the activities may be. Later on, as children reach 4 1/2 or 5 years of age, they often begin to ask why they come to occupational therapy, and at that point they can begin to understand the purpose of the activities. Most children have a desire to please adults, and many clients in the early stages of treatment find it easier to cooperate with their therapist to perform challenging fine-motor activities than with their parents! Therefore, the occupational therapist is often the first person who can coax a child into attempting something difficult, and the ability to grade activities and task analyze them helps occupational therapists to ensure successful experiences the first time. Occupational therapists have the ability to change the child's *attitude*, which may be the most important thing therapists can do to help a child. An important part of the art of therapy is to provide the client with enough support to want to risk something new or difficult.

It is important for the child to establish a good working relationship with the occupational therapist before activities are introduced at home, so therefore, fine-motor "homework" is not usually assigned in the initial months of therapy. Also, it is often difficult for parents to adopt a low-pressure, encouraging attitude with their child, due to their close relationship with the child. Sometimes the parent-child "fit" will not comfortably allow for a continuation of the therapy work at home. Decisions about whether or not to provide home activities are made individually for each child, depending on the unique family features of each specific case. Once therapy has begun, parents and teachers often begin to report a change in the child's attitude before they report a change in skill level. This is an important sign that the therapeutic process has begun. Comments such as "He is now choosing the fine-motor table at school," or "She started trying to paint when her big sister was painting a project at home," are often heard after a month or two of intervention. The willingness to try is the most important aspect of development that an occupational therapist can encourage. Once children begin enlarging their experiences of fine-motor challenges, their skill levels begin to improve and children often begin to bring in projects from school or from home. The Collaborative always has an ongoing art gallery in the occupational therapy room for such projects—at least for those instances in which the children will agree to part with them! The real "treatment," therefore, is an ongoing process throughout the week, and does not occur only during a therapy session.

The therapy session provides guidance and support for the child, and also encouragement for the parent during the frequent mini-consultations that occur as parents pick up their child after the treatment session. The therapists often have two main recommendations for parents. First, parents are encouraged to borrow or purchase tabletop easels for home activities since few of the parents are willing, due to realistic space limitations in their homes, to keep a free-standing easel up for more than a few days. The tabletop easels are smaller, and are more likely to be kept out for activities. Once parents understand the importance of working on a vertical surface, most are willing to purchase or make an easel for their child since it is such a simple adaptation to incorporate. Providing an easel as the routine setting for a variety of activities such as those suggested in this chapter usually does not seem like "homework" to the child or the parent, and can often be successfully incorporated at home very early on in the treatment process.

The second suggestion for parents is to purchase good-quality scissors for their child and to help teach their child to use the correct scissors grasp. Many parents also purchase a pair of identical scissors to leave at their child's nursery school.

Providing a vertical surface for activities and having the child use the correct scissors grasp are the two main areas in which parents can support treatment. However, some parents wish to provide more for their child, and those parents benefit from a discussion of toy selection and the use of therapeutic manipulative activities to the extent that it seems appropriate. Specifically, parents learn to understand the importance of manipulatives rather than writing utensils in promoting hand development. Parents are often asked to remove writing utensils from the child's daily array of activities, and instead to substitute some fine-motor toys and activities suggested by the occupational therapist. In particular, parents are encouraged to look at commercial toys in new ways. Many commercial toys "do it all" for the child, such as snowball-making tongs, bead-stringing machines, and so on. Other toys, such as games with small parts, tiny blocks, and miniature doll dishes, ask for skilled finger positions and regulation of the intrinsic muscles that are needed for skilled grasp and placement. Parents are asked to evaluate their child's toys and to work toward a balance between the toys that require little skill and the toys that require more skill. Parents learn that although a toy requires the use of the hands, such as

Legos building pieces, it does not necessarily require finger movements as much as wrist and arm movements, and therefore may not promote the development of fine-motor skill.

The ability to analyze the components of both therapeutic and day-to-day activities is one of the most important skills that occupational therapists have at their disposal. Although it would be impractical to fully teach parents and teachers this skill, it is possible to teach them to analyze fine-motor activities with a fair degree of understanding so that they are truly part of a team with the therapist. An involved parent can make important contributions, because once parents understand the concepts behind fine-motor development they are able to see activities in a different way. The parents and teachers feel empowered, and instead of feeling mystified or in awe of the therapist's special activities, they become contributors in an ongoing process. This kind of partnership does not diminish people's opinion of the occupational therapist's craft or abilities; rather, it strengthens respect and enhances the child's progress. It cannot be emphasized enough how important it is for everyone to *understand* the sequence of normal development, even if they are not taking an active part in providing the activities.

## Acknowledgments

I am grateful to my coworker and occupational therapist, Cindy Broder, for both inspiration and activities ideas. Furthermore, the children who attend our preschool collaborative program deserve recognition for their contributions of many of the wonderful ideas that have been included in this chapter. Additionally, the teachers in our preschool as well as those in area preschools, and the parents of the children, have also contributed many of the ideas. I would also like to thank Richard Myers for help with proofreading and editing, and for the illustrations.

## References

Benbow, M. (1987). *Sensory and motor measurements of dynamic tripod skill.* Unpublished master thesis. Boston, MA: Boston University.

Benbow, M. (1990a). *Developmental hand skill observations of the "K & l" child.* Distributed with handouts at a workshop, March, 1990.

Benbow, M. (1990b). *Loops and other groups, a kinesthetic writing system.* Tucson, AZ: Therapy Skill Builders.

Benbow, M. (1990c). *A neurodevelopmental approach to teaching handwriting.* Lecture notes from a workshop presented March 1990, and personal communication April 16, 1990.

Benbow, M. (1990d). *Understanding the hand from inside out.* Handout distributed at a workshop, March, 1990.

Connolly, K. (1973). Factors influencing the learning of manual skill in young children. In R.A. Hinde & J. Stevenson-Hinde (Eds.), *Constraints on learning.* London: Academic Press.

Gibson, R., & Tyler, J. (1989). *You and your child and playdough: Lots of play ideas for young children.* Tulsa: EDC Publishing.

Kapandji, I.A. (1982). *The physiology of the joints* (vol. I). New York: Churchill-Livingstone.

Long, C., Conrad, P.W., Hall, E.A., & Furler, S. (1970). Intrinsic-extrinsic muscle control of the hand in power and precision handling. *Journal of Bone and Joint Surgery, 52A,* 853-857.

Rosenbloom, L., & Horton, M.E. (1971). The maturation of fine prehension in young children. *Developmental Medicine and Child Neurology, 13,* 3-8.

Schneck, C., & Henderson, A. (1990). Descriptive analysis of the developmental progression of grip position for pencil and crayon control in nondysfunctional children. *The American Journal of Occupational Therapy, 10,* 893-900.

Wolfe, R. (1987). *Learn to cut.* Tucson, AZ: Communication Skill Builders.

# Appendix

## *Easel and Accessories Information*

Figure 10 shows a *tabletop easel* designed for a variety of uses. Paper can be attached for drawing and writing, and there is a ledge on which one can place Magna-Doodle, pegboards, geoboards, and so on. It also has a feature that keeps it from sliding across the table while being used, as well as storage space for art materials. It is painted to protect the surface from crayon marks.

Therapro, Inc.
225 Arlington Street
Framingham, Massachusetts 01701
(508) 872-9494

**Figure 10. Tabletop Easel from Therapro**

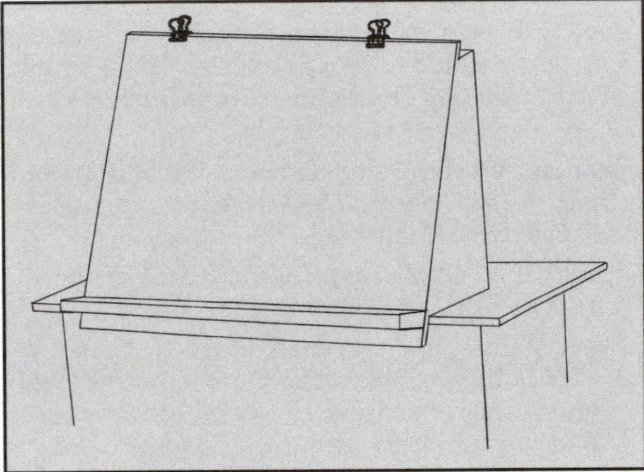

*Homemade easel:* This easel has been a very popular low-cost alternative for many schools and families. Art materials can be stored inside the box, and can help weight it down so it doesn't slide across the table when used (see Figure 11).

**Figure 11. Directions for Homemade Table Easel***

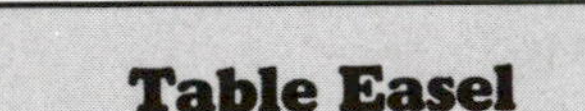

**Table Easel**

If you don't have an easel, it's easy to make one out of a large corrugated cardboard box. Just follow the diagram and these directions:

1. Have a grown-up help you make diagonal cuts along two sides of the box.
2. Fold flaps A and B in behind side C.
3. Now tape the bottom piece securely against the sides.

You could also use a plywood or Masonite board (sometimes called pressed wood) for an easel. Prop it up against the seat back of a kitchen chair. Then cover the seat with newspaper or plastic to hold a tray of paints, brushes, and water.

For paper, buy large pads of newsprint or save and cut up your large grocery bags — they're free! Use tape, thumbtacks, metal clips, or clothespins to attach the paper to the easel. Now you're ready to paint.

*Reprinted with permission from: Cole, A., Haas, C., & Weinberger, B. (1982). *Purple cow to the rescue* (p. 79). Boston: Little, Brown. Illustrations copyright 1982 by True Kelley.

*ChildWood learning units:* magnetic wooden figures and curriculum for early childhood education; these magnets are available in different themes and with familiar story figures, and you can also purchase a magnet story board that provides a vertical surface for using the magnets.

ChildWood, Inc.
8873 Woodbank Drive
Bainbridge Island, Washington 98110
(206) 842-3472

## *Manipulatives*

*You and Your Child and Playdough: Lots of Play Ideas for Young Children*, by Ray Gibson and Jenny Tyler. This colorful activity book includes many irresistible ideas for playdough projects, as well as several different recipes for making playdough. There are written instructions as well as simple picture instructions.

EDC Publishing
10302 E. 55th Place
Tulsa, Oklahoma 74146

*Waterfuls:* This toy is manufactured by the Tomy Corporation. The Waterful most recommended for Collaborative preschoolers is the full-sized Ring Toss toy, because it has continued to be the most appealing to a majority of children, and because it is one of the easiest Waterfuls with which a child can be successful. It has one button on the right side of the base. For left-handed children, or for a bilateral activity, the best Waterful is the new Ninja Turtles "Sewer Ball" Waterful with two buttons, one on each side.

Tomy Corporation, a subsidiary of Coleco Industries, Inc.
PO Box 580
Amsterdam, New York 12010

*Lite-Brite:* This is an electrically lighted pegboard that is oriented on a near-vertical slant. Because it has to be plugged in, the company recommends it for use with 4-year-olds and up. You can use premade picture designs to guide the peg placement, and picture refills and additional pegs are available. This item is often found at large toy stores such as Toys R Us.

Item #4780

Milton Bradley Company

Springfield, Massachusetts 01101

### *Scissors Information*

*Children's scissors:* These scissors were designed for children by Mary Benbow, OTR/L, and have been used very successfully in a variety of preschool and school settings. They are available in both right- and left-handed models, and are used in all of the classrooms of the Brookline-Newton Collaborative.

Item #120

OT Ideas, Inc.

111 Shady Lane

Randolph, New Jersey 07869

(201) 895-3622

*Learn to Cut: A Structured Program of Cutting Tasks with Reproducible Patterns*, by Robin R. Wolfe. This developmentally sequenced program teaches cutting skills and shape recognition simultaneously. In Part 1, eight types of cutting skills are presented in a hierarchy of steps. For each task, individual education program (IEP) objectives are stated and pre- and posttests, worksheets, and recording sheets are given. Part 2 contains patterns for 61 individual art projects, including 25 holiday cutting activities. All materials are reproducible.

Communication Skill Builders

3830 E. Bellevue/PO Box 42050

Tucson, Arizona 85733

(602) 323-7500

### *Drawing and Writing Materials*

*Preschool crayons:* The preschool crayons with the most therapeutic shape and the best quality are manufactured by the Rose Art Company.

Rose Art Industries

Bloomfield, New Jersey 07003

*Crayola primary-sized washable markers:* Labelled as "Broad Line Markers" and available in many toy and drug stores.

Crayola Consumer Affairs

PO Box 431

Easton, Pennsylvania 18044-0431

1 (800) CRAYOLA (weekdays between 9:00 and 4:00 pm EST)

*Magna-Doodle Drawing Board:* manufactured by Tyco Company and available at many large toy stores such as Toys R Us. Many accessories, including a variety of different magnet stamps (animals, circus) and tracing activities, are available. MagnaDoodle also comes in a smaller travel-sized version.

*Sanford's erasable plastic markers:* These crayons will fit into a Stetro pencil grip.

Item #89274

Sanford Corporation

Bellwood, Illinois 60104

(Also available from Therapro, see Easel section)

*Stetro pencil grips:* These are the most successful pencil grips we have found. They are sometimes available from educational supply companies or educational toy stores, or you can order them directly from Therapro (see Easel section).

# 5

# HANDWRITING: EVALUATION AND INTERVENTION IN SCHOOL SETTINGS

*Susan J. Cunningham Amundson*

Learning to write legibly is a major occupation of childhood. This intricate and complex process is one of the child's first tasks in an academic setting. Any youngster beginning to learn letter forms in the handwriting process is expected to have the requisite skills of organization, differentiation, sequencing, and memory (Kirk & Chalfant, 1984), along with postural control (Boehme, 1988) and eye-hand coordination (Getman, 1985; Laszlo & Bairstow, 1984). Despite the diverse readiness tasks needed for writing, many children by the age of 6 or 7 years, via a traditional instructional approach, are fairly proficient at writing in the school setting (Bergman & McLaughlin, 1988). Students with learning disabilities, developmental delays, or neurological impairments often struggle to write legibly for years when solely guided by the standard handwriting curriculum within the regular and special education classrooms. This incapacity and the end product of illegible handwriting directly affect the student's functional performance in academics (Hagin, 1983), self-esteem, and peer acceptance (Laszlo & Bairstow).

Despite the magic of word processors and augmentative communication systems, handwriting remains a necessary component of everyday life. Rosa A. Hagin (1983, p. 266) stresses the significance of handwriting in the educational setting:

> For pupils it represents the usual medium by which they convey to their teachers the progress they have made in learning what is being taught. Legible writing is a tool for learning; poor writing is a barrier. Unreadable numbers interfere with correct solutions in mathematics. Incomplete notes leave gaps in understanding in the content areas. Illegible examination papers bias test grades.

Since the implementation of Public Law 94-142 in 1975 (the Education for All Handicapped Children Act), occupational therapy as a related service in the educational setting has become more well-known and accepted by educators, parents, and school personnel. With the increased awareness of the profession's role related to students with perceptual motor deficits and a higher number of children labelled learning disabled than ever before, an upswing of referrals regarding children's poor handwriting has been reported by school therapists. Referrals from educators and parents frequently include that the child's handwriting is not readable, his or her writing is laborious and distorted, or he or she does not complete written assignments in class.

Unfortunately, children with handwriting dysfunction, when compared to peers who are experiencing moderate to severe disabilities, are often overlooked for occupational therapy intervention in educational settings. Low prioritization of "handwriting" students on occupational therapy caseload lists; lack of assessment tools, evaluation procedures, and intervention strategies for children's handwriting dysfunction; along with therapists' hesitancies to enter into an academic area, that is, writing, have all contributed to the impotency of the role of occupational therapy in this area. Conversely, occupational therapists possess the skills and expertise to make important contributions to interventions for handwriting dysfunction. An extensive neuromuscular and sensorimotor background, experience with functional training, knowledge about social and psychological behavior, competence in analyzing complex activities, and the capacity for making the most boring task enjoyable are all attributes that empower the occupational therapist to evaluate and expertly treat children with handwriting problems.

This chapter discusses the comprehensive evaluation process and effective intervention strategies an occupational therapist may use with a child with handwriting dysfunction in the school setting. Children with learning disabilities, emotional disturbances, and mild developmental disabilities will be the particular focus of its discourse. Specific areas addressed in assessment will include: examining underlying sensorimotor foundations of handwriting and the child's handwriting; briefly reviewing commercially available handwriting tests; and describing the process of gathering data related to the child's academic performance, cognition, psychosocial behaviors, handwriting curriculum, teacher's expectations, and parental support and involvement. Topics in handwriting intervention comprise discussing principles, research findings, approaches, and techniques of successful intervention programs and unifying these strategies with models of service provision.

## Handwriting Assessment Process

In the school setting, once a student with poor handwriting has been referred to occupational therapy by a team member, the assessment process may begin. A comprehensive handwriting assessment must include information and data gathered from classroom observation; teacher, parent, and other team member interviews; and formal and informal test instruments. Each child with handwriting difficulty looks different from every other child and only through the an individual assessment can therapists understand the sensorimotor foundations, cognitive and psychosocial behaviors, environmental factors, and skill acquisition that contribute to the child's failure to demonstrate legible handwriting.

### Classroom Observation

Skilled observation of the pupil writing in the classroom is mandatory. The examiner should consider the child's behavior and whether or not the child can attend to the task of writing independently. Other concerns address the child's distractibility—visual or auditory—and the child's level of frustration with writing. During the observation, the difficulty level of the writing task should also be noted. For example, a writing task of composing a sentence requires different sensorimotor and cognitive skills than dictating or copying a sentence. The classroom environment, the child's actual physical placement in the room, and the child's interaction with the teacher and peers also lend more information to the therapist about the child with handwriting difficulties.

### Interviews

Interviewing the educator, parents, and other team members is another powerful ingredient in gathering data about the referred student (Cook, 1991). The educator can describe the child's strengths and areas of concern in his or her academics, state how the child's performance in writing and other subjects compares to his or her peers, and explain the specific handwriting curriculum adopted by the school and the writing proficiency expected at the particular grade level. Handwriting curricula may vary from school district to school district, from school to school, and from grade to grade. Expectations of educators may also differ regarding writing proficiency at grade levels. These variables may heighten the confusion for the child already struggling to write legibly if he or she transfers between districts, schools, grades, and teachers.

Parents can also provide invaluable information related to their child in the assessment process. Serving as critical team members, parents can provide information related to the child's medical, familial, developmental, and play histories (King-Thomas & Hacker, 1987). Family expectations of the child related to writing can also be revealed in the interview. Does the family encourage or require the child to write his or her name, complete written homework assignments, record telephone messages, or correspond with relatives and friends in the home environment? Their perceptions about the child's interests, self-concept, and social conduct may also be communicated to the therapist. The interviews with the parent and team members have a twofold purpose as they serve as mechanisms for gathering information and building rapport (Cook, 1991).

### Informal and Formal Tests

Test instruments allow therapists to investigate many facets of children's handwriting including the underlying sensorimotor foundations, cognitive and psychosocial behaviors, written product, and ergonomic factors. Whether the test be norm-referenced, criterion-referenced, or informal, the findings can uncover the specific areas of concern as well as the child's strengths related to handwriting and can direct the occupational therapist, the parents, and educational team in planning an effective intervention program.

## Evaluating Handwriting

### Underlying Sensorimotor Foundations

Previous chapters have addressed certain aspects of sensorimotor functioning and hand development.

Therefore, this one will briefly review the sensorimotor components comprised of neuromuscular mechanisms, sensory integrative functions, and motor control and the relationship of these components to their roles in the functional performance of handwriting.

### *Neuromuscular Mechanisms*

Neuromuscular mechanisms include postural control, upper-extremity stability, and muscle tone. *Postural control* is the base of stability from which any purposeful movement may occur (Boehme, 1988). Frequently, children lacking a stable trunk and experiencing variations and fluctuations in *muscle tone* demonstrate poor handwriting. They are unable to sustain an upright position and make the needed postural adjustments while focusing on fine-motor activities—for example, handwriting (Price, 1986). Another neuromuscular requisite, *upper-extremity stability,* particularly through the shoulder girdle, may be affected by variations in muscle tone, an inadequacy to cocontract agonistic and antagonistic muscle groups around joints, and lack of fluidity of scapular movements. Figure 1 displays an occupational therapist evaluating a student's proximal joint stabilization.

Without the stabilization of the shoulder, elbow, and wrist, the speed and dexterity of the hand's intrinsic movements when manipulating the writing tool become impeded. Insufficient neuromuscular mechanisms of children commonly interfere with legible handwriting.

### *Sensory Integrative Functioning*

The areas of sensory integrative functioning most directly related to handwriting encompass the tactile, proprioceptive, and visual systems; kinesthesia; visual perception; and motor planning. The *tactile* and *proprioceptive systems* provide information to the child regarding the grasp of pencil and eraser, and the stabilization of the paper. *Visual demands* for writing tap heavily into the oculomotor system mediated by the vestibular system (Dunn & Lane, 1986). Maintaining the position of head and eyes, saccadic eye movements, and smooth eye pursuits affect a child's ability to sustain visual regard, scan a line of print, and focus on stationary and moving targets, such as one's own pencil forming numbers. Influencing the amount of pressure a child applies to the writing tool and the paper, the ability to write within perimeters, and the directionality of the implement is *kinesthesia,* which is the awareness of the extent, weight, and direction of movement. Children with kinesthetic dysfunction may press too hard or too softly with their pencil while writing and may be confused when needing to direct their writing tools to form letters or to write between lines. *Visual perception*, the ability to organize and interpret what is seen, affects handwriting chiefly in two areas. Position in space or the ability to "determine the spatial relationship of figures and objects to self or other forms and objects" (McGourty et al., 1989) affects a child's ability to space between letters of a word and words of a sentence and to place letters correctly on the writing line (horizontal alignment). Commonly, children with poor position in space, when writing, may demonstrate underspacing (running together or overlapping letters and words), overspacing (leaving large spaces between letters and words) (Stott, Moyes, & Henderson, 1985), and improper placement of letters on the writing line. Form constancy, the capacity to "recognize forms and objects as the same in various environments, positions, and sizes" (McGourty et al.) allows a child to discriminate between letters and numerals that are finitely similar, such as, *b/d, p/q, 2/5, S/S,* which is critical for writing. The lack of form constancy may result in reversals of letters and transpositions of words. Finally, *motor planning* in writing influences the child's ability to plan, sequence, and execute letter forms and the ordering of letters to build words.

**Figure 1. The Therapist is Evaluating the Child's Proximal Joint Stability Along with His Scapular Rotation in the Quadruped Position**

### *Motor Control*

Motor aspects related to handwriting involve activity tolerance, bilateral integration, visuomotor integration, and fine-motor coordination. Obviously, *activity tolerance*, sustaining a writing activity for a duration of time, is needed in order for the child to practice and master the skill. Both strength and endurance are needed to persist in a handwriting task for an extended period of time. *Bilateral integration* includes both the symmetrical and asymmetrical movement the body needed for an activity (Exner, 1989). Writing consists mainly of asymmetrical movements—for example, stabilizing the paper with the nonpreferred hand while holding the pencil with the preferred hand—and may not be achieved when children have difficulties dissociating movements of the upper extremities. *Visuomotor integration,* the ability to coordinate visual information with a motor response, allows a child to reproduce letters and numbers for written school assignments.

According to Exner (1989), three aspects of *fine-motor control* that affect handwriting are: (1) isolation of movements, (2) grading of movements, and (3) timing of movements. An inability to *isolate* and *grade* finger and hand movements may result in inadequate pencil grasp along with an inability to monitor discrete movements (Exner, 1989). Children demonstrating poor grading of movements usually use compensatory techniques to stabilize their pencils, most frequently observed as fingers locked into extension or fisted into flexion. *Timing* affects the rhythm and flow of writing; inadequacies are exhibited by the child as slow, labored, jerky writing or rapid, haphazard penmanship.

Fine-motor control includes *in-hand manipulation*, described by Exner (1989, p. 242), as "the process of adjusting objects within the hand after grasp." After grasping a pencil, "shifting" (the linear movement of the tool among the fingers needed to adjust it for writing) and "rotating" (the movement of the tool around an axis) are essential for vertically turning the pencil from grasp to placement for writing or for erasing (Exner; Boehme, 1988).

## Cognitive and Psychosocial Behaviors

Cognitive and psychosocial information contribute to the overall picture of the child with handwriting dysfunction. Occupational therapists need to consider the child's attention span, memory (visual, verbal, and motor), sequencing, and conceptual skills during the assessment. These factors along with the child's self-concept, interests, behaviors, and motivation affect his or her performance of handwriting and should be incorporated as facets of the evaluation. Often, children with poor handwriting develop low self-esteem. Each time the child writes in a sloppy, haphazard manner, it looks right back up at him or her and states, "This is a mess. You are such a messy kid." For the child, this message is a constant reminder of failure at school, contributing to an already fragile self-esteem. From this poor self-concept, acting-out behaviors and low motivation for writing may result.

## Handwriting Assessments

### *Commercially Available Assessments*

Several assessments to measure the handwritten product and writing mechanics of a child are commercially available to occupational therapists. The most commonly used ones in the United States include the *Children's Handwriting Evaluation Scale* (Phelps, Stempel, & Speck, 1984), the *Children's Handwriting Evaluation Scale-Manuscript* (Phelps & Stempel, 1987), the *Denver Handwriting Analysis* (Anderson, 1983), the *Diagnosis and Remediation of Handwriting Problems* (Stott, Moyes, & Henderson, 1985), and the *Test Of Legible Handwriting* (Larsen & Hammill, 1989). Table 1 contains a summary of the features of these tests.

The occupational therapist should select the instrument that best addresses the concerns of the child's handwriting and that guides program planning. The therapist should, at the same time, keep in mind the test's psychometric properties, e.g., its reliability and validity.

### *Measuring Areas of Handwriting*

In the educational setting, whether using formal or informal tests to measure children's handwriting, domains of handwriting, legibility components, speed of writing, and ergonomic factors must be carefully examined. Handwriting tasks similar to those required of the student in the classroom reveal specifically what troublesome areas a student may be experiencing. The approach to and execution of dysfunctional writing must be carefully observed by the examiner to determine which legibility components are inadequate and are contributing to the illegible written product. Finally, the speed of writing must be regarded with respect to written assignments in the classroom.

*Domains of handwriting,* the tasks required of students in the classroom, including dictation, far-point copying, near-point copying, manuscript-to-cursive transition, upper-and lower-case letter writing, composition, and endurance, must be targeted in the assessment. *Dictation* is the writing of letters or words when verbally requested by the examiner and emphasizes the child's ability to motorically respond to auditory directions. This skill is particularly important in spelling words. *Copying* is reproducing letters, numbers, and words from a model either from manuscript to manuscript or cursive to cursive. Frequently, copying tasks occur in the classroom as children copy mathematical problems from textbooks and as they reproduce a sentence from a model on a chalkboard while seated at their desks. Far-point copying entails copying from a distant vertical model to the horizontal writing surface, whereas near-point copying is duplicating letters or words from a nearby horizontal surface, usually on the same page or at least on the same writing surface. *Manuscript-to-cursive transition* is transcribing manuscript letters and words to cursive letters and words and obviously requires a command of both manuscript and cursive letter forms. *Writing the alphabet in both upper- and*

**Table 1. Handwriting Assessment Tools**

| | Children's Handwriting Evaluation Scale | Children's Handwriting Evaluation Scale Manuscript | Denver Handwriting Analysis | Diagnosis and Remediation of Handwriting Problems | Test of Legible Handwriting |
|---|---|---|---|---|---|
| **Age or Grade Range** | **Grades 3–8** | **Grades 1–2** | **Grades 3–8** | **None** | **7–18.5 Years** |
| **Script assessed:** | | | | | |
| Manuscript | | X | | X | X |
| Cursive | X | | X | X | X |
| **Domains tested:** | | | | | |
| Composition | | | | X | X |
| Dictation | | | X | | |
| Endurance | | | X | X | X |
| Far-point copying | | | X | | |
| Manuscript to cursive transition | | | X | | |
| Near-point copying | X | X | X | | |
| Upper-lower case alphabet | | | X | | |
| **Type of test:** | | | | | |
| Norm-referenced | X | X | | | X |
| Criterion-referenced | | | X | X | |
| **Scores obtained:** | | | | | |
| Percentile | X | X | | | X |
| Standard | | | | | X |
| Quantified observation | | | X | X | |

*lower-case letters* involves the child's sequencing of the alphabet, the consistency of letter cases, and the formation of each individual letter. *Composition* requires the student to compose a sentence or paragraph and involves an integration and synthesis of cognitive skills more complex than those needed for a simpler writing task, such as copying. As children create stories and poems for assignments, they demonstrate composition. By testing as many of the domains of handwriting as possible, the occupational therapist can glean which specific tasks the child may be struggling with in the classroom and develop an intervention program that addresses them.

When examining a child's handwriting, the first question the examiner may ask is "Is it legible?" Ziviani and Elkins (1984) briefly traced the history of handwriting *legibility* and found that, in 1912, legibility was judged by the time taken to read a child's handwriting sample. However, researchers in the past 2 decades have judged legibility in terms of letter formation, alignment, spacing, size, and slant (Ziviani & Elkins). Any one of these components acting alone may affect the readability of handwriting, but when two or more are present, the legibility of the writing sample may be drastically reduced.

Alston and Taylor (1987) identified significant contributions to handwriting illegibility by analyzing 100 handwriting samples of 7- and 8-year-old children in England. In *letter formation,* she found the most common characteristics included: (1) incorrect letter forms, (2) inadequate "leading in" and "leading out" of letters, (3) poor rounding of letters, (4) incomplete letter closure, and (5) inadequate letter ascenders and descenders.

Other components also contribute to illegibility. *Alignment* refers to the placement of the letter on the writing line along with the ability to write on lines of appropriate width for age expectations. *Spacing* is "the way letters are distributed within words and how words are spaced within sentences" (Larsen &

Hammill, 1989, p. 21). Uniformity of spacing is key, with the recommendation that spacing between words be slightly more than the width of one lower-case letter (Larsen & Hammill). Determining if letters are similar in *size* and if the *size* of writing is appropriate also contribute to legibility. Although difficult to measure objectively, *slant* should be consistent, and when it is not uniform, it makes handwriting difficult to read even when other components of legibility are intact. Figure 2 displays faults of legibility components.

Research studies investigating children's writing *speed,* or the number of letters written per minute, have not resulted in congruent information. In the studies reviewed (Ziviani & Elkins, 1984), the range of writing speed at Grade 5 is from 38.4 letters per minute to 64.0 letters per minute with the variability possibly due to different methodologies, subjects being from various countries, and data collection occurring in four separate decades. Currently, baseline data are vague, but for students to accomplish an acceptable amount of work in the classroom, a certain writing speed must be maintained. The speed and level of skill in writing may deteriorate when the complexity or volume of the writing task increases or the speed demanded is faster than the child's natural writing speed (Rubin & Henderson, 1982). Therefore, the occupational therapist and teacher must determine if the child's written productivity is adequate for the time constraints and volume of work involved.

*Endurance* is observed in testing or in the classroom while the child writes five or six sentences. Poor handwriters frequently cannot sustain the legibility of their writing as the length of their assignment increases. As the student tires, letter sizes may become smaller, writing more labored, and letters and words omitted. When endurance is poor and the student's writing pace is slow, he or she does not have the opportunities to practice letter formation as do other children. In these children, mastery of letter formation may be poor due to insufficient practice (Anderson, 1983).

As children are engaged in the writing tasks of the assessment, *ergonomic factors* such as posture, upper-extremity stability and mobility, and pencil grasp must be carefully observed and documented. The child's sitting posture is analyzed. Does the child support his or her head when writing or lie the head on the desktop? Does the child kneel in the chair to gain more proprioceptive input through the arms? Are the chair and desk at appropriate heights for writing in the classroom? *Stability and mobility of the upper extremities* refers to the stabilization of the shoulder, elbow, and wrist to allow for distal mobility

**Figure 2. Components of Handwriting Legibility**

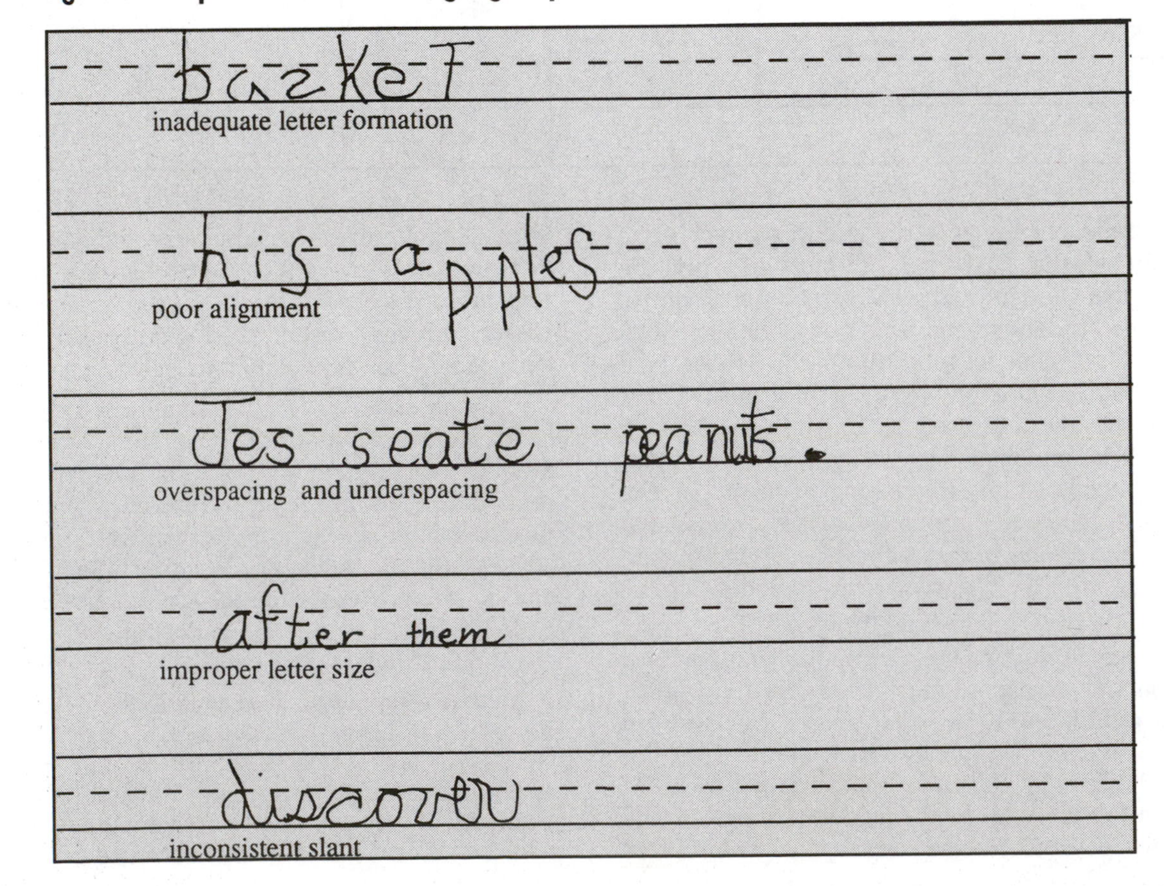

of the hand to manipulate the writing tool. What is the position of the writing arm? What other parts of the body are moving during writing? Are associated reactions present? Is appropriate fixation and release occurring at the elbow and wrist joints of the writing hand? Does the assisting hand steady the writing paper? Is the writing movement smooth or jerky? Finally, the primary ergonomic focus for most occupational therapists is *pencil grasp*.

How does the child specifically grip the pencil? Is the thumb pad opposed to the pad of the index finger and supported by the radial side of the middle finger? What amount of pressure on the pencil is used? When directed, can the student adjust the pencil in the hand?

The dynamic tripod grasp involves resting the writing tool on the distal phalanx of the radial side of the middle finger while controlling it between the pads of the thumb and the index finger (Rosenbloom & Horton, 1971). This is typically the grip most occupational therapists and educators encourage children to use when writing (Myers, this volume). Although the ideal dynamic tripod posture has been strongly advocated, variations are typical (Schneck & Henderson, 1990; Ziviani, 1983; Ziviani, 1987) among children without handwriting dysfunction.

Schneck and Henderson (1990) studied 320 nondysfunctional children, ages 3 years, 0 months to 6 years, 11 months, to examine grip position for pencil and crayon control. In their investigation of children drawing and coloring, they observed 10 developmental grips and classified them as primitive, transitional, or mature. Mature grips were defined as the dynamic tripod grasp and the lateral tripod grasp. See Figure 3 for illustrations of these grasps.

Their findings indicated that either the dynamic tripod or lateral tripod grasp was demonstrated by 95% of the children aged 6 1/2 to 7 years when performing a drawing task. Approximately one-quarter of children ages 5 years, 0 months to 6 years, 11 months, preferred the lateral tripod grasp, contrary to the traditional stance of educators and occupational therapists that children should be using the dynamic tripod grasp for writing. This study provides important baseline data about children's pencil and crayon grips and indicates the normal variability of grasps among nondysfunctional children. In the assessment process, occupational therapists, when observing pencil grasp, must carefully describe and classify grips to compare and contrast them with the normal variations that have been documented in the literature.

**Figure 3. Mature Pencil Grasp*—(a) Lateral Tripod Grasp (Schneck, 1991); (b) Dynamic Tripod Grasp (Rosenbloom & Horton, 1971)**

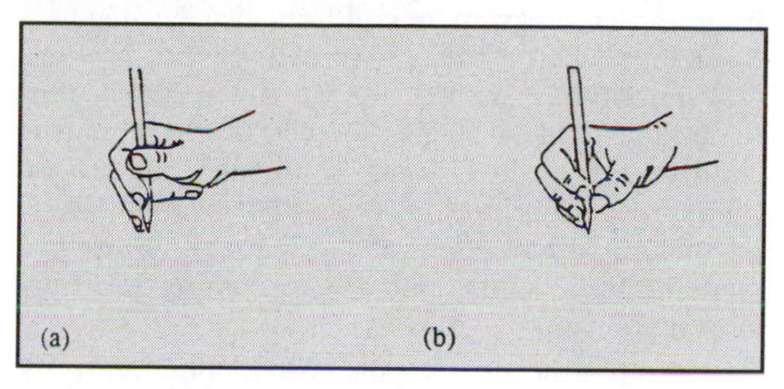


*Illustrations reprinted with permission from C. Schneck.

**Figure 4. Therapeutic Approaches of Handwriting Intervention**

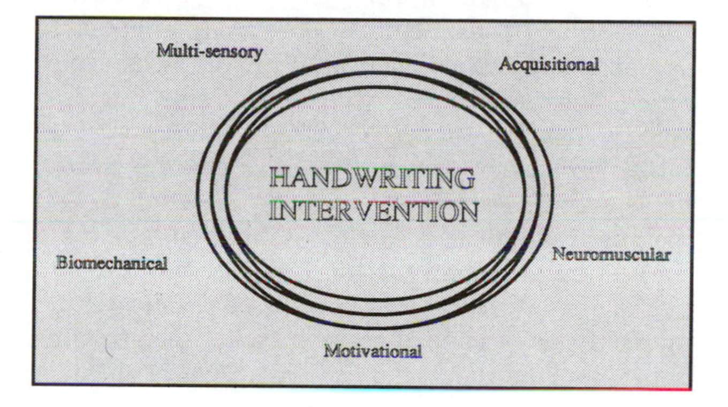


## Handwriting Intervention

From a traditional, educational model, occupational therapy intervention with a child with handwriting dysfunction may appear unique. Occupational therapists possess the power to use a variety of therapeutic approaches for remediation of handwriting problems. Theoretical approaches that apply to handwriting intervention include: (1) neuromuscular, (2) acquisitional, (3) sensorimotor, (4) biomechanical, and (5) motivational (see Figure 4).

When considering any intervention plan for handwriting, therapists must keep in mind all five approaches and the dynamic balance between them (Holm, 1986). At certain times in a therapy session or a progression of therapy sessions, one approach may recede as another receives more emphasis. However, each one needs to be held in equilibrium with the others, with therapists always cognizant of the meaningful end goals for the student. By holding these theoretical approaches in balance, while planning and implementing treatment with a child with handwriting dysfunction, the occupational therapist opens avenues of learning and increasing opportunities for the child to master.

### Neuromuscular Approach

Postural preparation to write originates from this construct. Preparing children's bodies and upper extremities to write is critical in any type of handwriting intervention program, whether it is delivered in the classroom among peers or in the isolated therapy room. Maddox (1986) suggests that any handwriting intervention program for children with developmental delays "should begin with attention to postural and muscle tone deficits" (p. 3). Her recommended techniques and strategies for posturally preparing children target three areas: (1) modulating muscle tone throughout the trunk and upper extremities, (2) ensuring free movement of the scapulothoracic and glenohumeral joints, and (3) promoting proximal joint stability.

Modulating muscle tone throughout the trunk may consist of increasing, decreasing, or balancing tone (Maddox, 1986). Increasing tone in children may be achieved by activities such as rolling up a gradual incline with hands held above one's head, jumping on a hippity hop ball, or spinning on a "sit-and-spin." Older children, particularly boys, may enjoy more aggressive activities such as jumping jacks; "commando" obstacle courses that include shinning up a pole or fast foot maneuvering of a tire pattern; or lifting and pushing furniture. For children whose muscle tone needs to be decreased, slow rolling and rocking are the most powerful inhibitory influences to the vestibular system (Oetter, 1986). In the classroom a child's postural tone may be reduced prior to writing by rocking in a padded rocking chair, keeping time with slow, rhythmical music from a headset or with a metronome set at a slow tempo, such as *adagio*. According to Maddox, balancing muscle tone can be achieved through active weight shifting; smooth, repetitive, alternating movement; and controlled rotation. A student needing more balanced tone might engage in shifting weight in a half-kneel stance or bending laterally to the left and right in sitting—both unobtrusive classroom activities for preparing to write.

Occupational therapists may also prepare children posturally for writing by ensuring free movement of the child's scapulothoracic and glenohumeral joints. Some prewriting activities may be performed by the child with the therapist's supervision; however, hands-on contact by the therapist may be needed to facilitate smooth movement. Humeral rotation may be facilitated through activities involving weight bearing, compression, and traction (Boehme, 1988; Maddox, 1986). In the classroom, child-activated exercises to improve scapular movements include having the seated child extend his or her arms forward with the backs of hands together, making slow, circular arm motions, or having the seated child, with hands on opposite shoulders, lift his or her elbows up and back down slowly (Maddox).

Frequently, children with handwriting dysfunction experience poor proximal joint stability. To encourage cocontraction through the neck, shoulders, elbows, and wrists, young children may enjoy animal walks requiring weight bearing on the upper extremities, such as the bear walk, the inchworm creep, or the mule kick (see Figure 5).

Other creative games involving the medicine ball, wrist weights, or large therapy ball can provide opportunities for more proximal joint stabilization. Older children might prefer performing wall push-ups, calisthenics, and theraband (or elastic, commercial fishing tube) resistive exercises in the school hallway to prepare themselves for writing.

When needed, postural preparation must be the preliminary component of handwriting intervention and should occur in a 5- to 10-minute period before the instructional writing program begins. The therapist will need to carefully select activities targeting the child's deficits in postural control, joint movement, and/or proximal joint stability. Other preparatory activities for in-hand manipulation, visual perception, motor planning, and related sensorimotor components may be necessary, depending on the student's needs; however, postural preparation should immediately precede the actual writing instruction.

**Figure 5. The Child is Engaged in a Fun Activity—the Mule Kick (an Example of Postural Preparation Prior to Handwriting)**

## Acquisitional Approach

In any handwriting intervention, a sequenced, instructional program is required to enable the child to acquire and master handwriting proficiency. This skill, like other acquisitional skills, "can be improved through practice, repetition, feedback, and reinforcement" (Holm, 1986, p. 70).

Graham and Miller (1980) purport principles and conditions fundamental to a successful and effective handwriting program, which occupational therapists should consider. They recommend handwriting instruction be: (1) taught directly and not incidentally; (2) implemented in short, daily lessons; (3) individualized to the needs of the student; (4) planned, monitored, and altered based on assessment data; and (5) overlearned and applied in a meaningful context for the student. By incorporating these conditions—along with stressing the importance of handwriting, creating a positive learning environment, and encouraging a consistent, legible writing style—therapists and educators will cultivate and encourage efficient, legible writers (Graham & Miller; Milone & Wasylyk, 1981).

A structured instructional program of letter and numeral formation (Taylor, 1985) must be implemented with letters being introduced in a sequential progression. Based on the assessment results, each student's program will be individualized because the letters already mastered can be omitted from the direct instruction. The program should build on mastered letters and those being learned, excluding any letters the child does not know or is forming improperly. To practice letters incorrectly is to rivet unwelcome perceptual-motor patterns in the child who needs no more reinforcing of poor handwriting (Ziviani, 1987).

As soon as two or three letters are introduced and mastered, the child can then begin to put them in a meaningful context, namely words. For example, if a child had only learned the lower case manuscript letters of *a* and *d*, the child could write two words, *add*, *dad*. Each intervention period, newly learned letters are joined with previously mastered ones to form words during the session. This immediate reinforcement of writing words is much more purposeful and powerful for the child than writing a string of repeated letters.

Reviews of handwriting research (Bergman & McLaughlin, 1988; Peck, Askov, & Fairchild, 1980) indicate that numerous approaches to remediate and instruct handwriting have been studied. However, a combination of techniques seems more desirable and effective than any one method in improving handwriting. The following instructional approach sequentially blends modeling, tracing, stimulus fading, copying, composing, self-monitoring, and peer recording and has been highly successful in teaching and improving handwriting in the academic setting. Recommended successive steps are:

1. *Modeling:* The therapist names the letter to be written. Then, he or she makes the letter on the lined writing surface while verbally cuing the child. The therapist not only models the letter but also models the action to write the letter, repeating it two or three times.
2. *Tracing:* The child traces over the letter, as the therapist repeats the verbal cues and if needed, provides the child visual cues also (e.g., colored dots for starting points) (Bergman & McLaughlin, 1988). When providing verbal cues, therapists need to consistently use direct and simple handwriting terms with the students (Milone & Wasylyk, 1981).
3. *Stimulus fading:* The verbal and visual cues are retained to reduce or circumvent errors for children but must be reduced as soon as the child can successfully form the letter without them.
4. *Copying:* Although some children can imitate a letter correctly after modelling, children with perceptual motor deficits may require both tracing and stimulus fading in order to reach the copying phase (Bergman & McLaughlin, 1988). When copying, the child does not observe the therapist producing the letter, but only the end product, and then attempts to reproduce it. Children should copy the learned letter until formation is acceptable to the therapist and the child. Once mastery of the letter is achieved, the child can begin copying words that include the learned letter and previously mastered letters. A level of acceptability should be determined that is both challenging yet achievable for the student (Milone & Wasylyk, 1987).
5. *Dictating:* Following successful copying, the therapist may dictate words and the spelling of words for the child to write.
6. *Composing:* Children may be able to conjure up their own words, which include mastered letters, for their writing group to practice. Subsequently, they begin to create sentences.
7. *Self-monitoring:* This technique requires each child to become more actively involved in the handwriting program by assuming responsibility for correcting his or her own work (Bergman & McLaughlin, 1988). With younger children, self-correction may occur verbally as they analyze letter formation and overall legibility, whereas the older child might follow a checklist to examine written work. Figure 6 displays a checklist for legibility used with older children in the self-correction process.

**Figure 6. Self-Correction Checklist for Older Children**

**Handwriting Check-up**

| | YES | NO |
|---|---|---|
| 1) My spacing is correct. | ___ | ___ |
| 2) My letters are the correct size. | ___ | ___ |
| 3) My letters are on the line. | ___ | ___ |
| 4) My letter forms are correct. | ___ | ___ |
| 5) My slant is in one direction. | ___ | ___ |

8. *Peer recording:* When instructing a group of children, peer recording or correcting one another's written product, is an influential tool, especially when children are unable to be objective in self-correcting. By evaluating a peer's handwriting, children are able to be more objective in analyzing and marking accuracies and faults, and then they can apply similar standards to their own written product. In this process, the therapist must be very supportive and model positive reinforcement for noting "good" writing as well as errors.

When learning proper letter formation, other components of legibility—size, slant, and alignment—are addressed inherently . Frequently, spacing and speed are not. Overspacing and underspacing between words is common among children with perceptual deficits. The space between words should be slightly more than the width of a single lower-case letter (Larsen & Hammill, 1989) and may be demonstrated by using actual objects of the necessary width (e.g., width of pencil shaft) to represent the "space bar," depending on the width of the writing lines. The object can later be imagined by the student rather than placed on the writing surface.

Legible writing is nonfunctional unless it can be produced in a timely manner (Alston & Taylor, 1987). One of the most common concerns related to students with handwriting dysfunction in the academic setting is that they are very slow and are unable to complete written assignments in a specified time (Rubin & Henderson, 1982). The reasons for this are twofold. First, children with poor mechanics for writing are deterred even further when needing to integrate spelling and composing into the process. Second, students with poor handwriting usually have an unstable motor set and are unable to write with automaticity and flexibility (Rubin & Henderson). Thus, overlearning of the legibility components is mandatory before speed can be stressed (Anderson, 1983; Bergman & McLaughlin, 1988). Even then, speed may not increase for children with severe perceptual-motor difficulties.

Classroom methods of handwriting instruction vary in the United States with the commonly used ones being Palmer, Zaner-Bloser, italics, and D'Nealian (Alston & Taylor, 1987; Duvall, 1985; Thurber, 1983). When designing a remedial program for a student, the educational team must endorse one handwriting method and one style (manuscript or cursive); otherwise, the child may become confused. A popular handwriting method in the school setting, D'Nealian possesses features that are advantageous for the poor handwriter. D'Nealian manuscript letters are formed with one continuous stroke. This encourages the development of rhythm at the onset of writing, and early learning is built on rather than unlearned during later cursive instruction (Thurber). Letters are presented in groups of similarly made construction, which allows opportunities for the child with perceptual-motor difficulties to discriminate immediately between them, visually and motorically. The method also includes using newly learned letters in words and learning lower-case letters before upper-case.

With regard to the handwriting method used, the educational team and the occupational therapist must select a style of writing (manuscript or cursive) most advantageous for the student with handwriting dysfunction. Sometimes, students may request to learn or practice one style over the other, and such a request also influences the team's decision. Traditionally, manuscript writing is accepted as the general approach for use in grades 1 and 2, with cursive writing introduced at the end of grade 2 or the beginning of grade 3 (Bergman & McLaughlin, 1988; Hagin, 1983). Educational research has not conclusively indicated the superiority of one style of writing (Graham & Miller, 1980; Hagin). Therefore, therapists and educators must consider the advantages and disadvantages of manuscript and cursive before they can decide which style would be most suitable and successful for the student requiring intervention.

Each style of writing has attractive features for the child with handwriting dysfunction. Manuscript is advocated for young children for the following reasons:

1. Simple letter forms make it easier to learn.
2. It resembles the print of books.
3. Beginning writing in manuscript is more legible than cursive.
4. It is required for applications and documents in adult life.
5. Ball and stick forms of manuscript letters are more developmentally appropriate than cursive letters for younger children (Barbe, Milone, & Wasylyk, 1983; Hagin, 1983).

Proponents of cursive writing state:

1. It allows the student to deal with words as units.
2. Individual letter forms are difficult to reverse and words are not easily transposed as in manuscript.
3. It is faster.
4. It offers a new form of writing (like grown-ups' writing) for the poor printer who may possess more maturity at the introduction of cursive writing (Armitage & Ratzlaff, 1985; Hagin, 1983).

## Multisensory Approach

Olsen (1980) suggests a multisensory approach for handwriting intervention and claims it is effective due to children's various learning styles and the complexities of the writing task. By varying sensory experiences, the child's nervous system may integrate information more efficiently to produce a satisfactory motor output (i.e., legible handwriting). Using a multisensory approach, students with perceptual-motor difficulties remain interested, challenged, and enthusiastic, in spite of previous failures in handwriting. All sensory systems can be tapped, including the olfactory, gustatory, visual, proprioceptive, tactile, and auditory senses, creating more vehicles of information arriving to the child's nervous system.

Occupational therapists incorporating a multisensory approach into handwriting intervention may use a plethora of instructional materials and methods. Most students who have been struggling to write legibly prior to occupational therapy intervention have been instructed at a seat and desk or table using a No. 2 pencil and age-appropriate writing paper. If children are unsuccessful with tedious pencil-and-paper tasks, other modalities must be tried. Writing utensils, writing surfaces, and positions for writing should all be considered in the handwriting intervention program.

Examples of writing tools include magic markers, felt-tip pens, crayons and "wipe-off" crayons, grease markers, weighted pens, erasable-ink pens, wooden dowels, and chalk. Small pieces of chalk, preferably colored, impel children to use a radial digital grasp when writing (Weiser, 1986). Lamme and Ayris (1983) investigated the effects of five writing tools on handwriting legibility of 798 1st-grade children. Their findings revealed that the type of writing tool did not influence legibility, and the wide "primary" pencil did not enhance legibility any more than the adult No. 2 pencil for beginning writers. However, educators involved in the study noted children's attitudes toward writing were more positive when they were able to use a felt-tip pen rather than a pencil. This suggests that a wider variety of writing tools might improve children's feelings about writing. Finally, an effective writing tool for older children learning cursive is the photoelectric pen or "talking pen."[1] This special light-sensitive tool provides feedback through a buzzing sound when the pen controlled by the student goes off-track while tracing over black cursive letter forms and is met with enthusiasm by many upper-elementary students.

Writing surfaces may be in vertical or horizontal planes. Regardless of axis, writing lines are necessary to serve as direction indicators, height controllers, and letter positioners (Pasternicki, 1987). Vertical surfaces may include the chalkboard and laminated paper sheets and poster paper attached to the wall (see Myers, this volume, for other examples of activities on vertical surfaces). An upright orientation may lessen directional confusion of letter formation for the student with perceptual-motor difficulties (Hagin, 1983). At the chalkboard, up means up and down means down, whereas on a horizontal surface, up means away from oneself and down means toward oneself. Writing on the chalkboard and laminated paper sheets also provides additional proprioceptive input as more pressure for writing is required than the traditional paper and pencil medium. Particularly effective is the laminated sheet that is color-coded. Below the solid writing line, the color brown represents the "ground" between the solid line; the dashed green guideline stands for the "grass;" and above the guideline to the top line, blue is for the "sky" (Oetter, 1986). For example, the letter *j* would start at the top of the grass and go down into the ground with the dot in the sky (see Figure 7). These concrete concepts are more understandable for the younger child than abstract directional concepts.

**Figure 7. Writing on Laminated, Color-Coded Paper Can Often Give Children with Sensorimotor Deficits More Cues for Proper Letter Formation**

[1]Available from Wayne Engineering, Winnetka, IL 60093.

Clay trays, plastic bags filled with gel, the chalkmat, and trays filled with Tang, dry pudding, sand, or shaving cream are motivating horizontal writing surfaces for children. Clay trays are cookie sheets, pizza pans, or Styrofoam meat packaging trays filled with modelling clay. With a dowel, the child can practice letter forms in a very resistive medium. The chalkmat, a portable, flexible chalkboard, can be used on the floor for children to write on and is particularly well-suited for the itinerant therapist. Other intriguing textured surfaces might be dry pudding mix or Tang in a shallow tray for students to use isolated fingers to practice letters.

Along with using a wide array of utensils and surfaces in intervention, children's positions while writing are critical. Having total body extension in a plane parallel to the writing surface seems to allow for more internal stability of the trunk and more proprioceptive input through the upper extremities. Therefore, students may benefit from standing at a chalkboard while writing or from lying down on a soft mat with a pillow tucked under their chest while practicing letter forms in a clay tray. Standing entails having both arms forward with slight elbow flexion as the bases of the palms rest on the vertical surface. This posture may increase trunk control, improve proximal joint stability, promote the hand crossing the body's midline, and allow a disassociation of hand movements from the forearm during writing activities. Lying prone encourages weight bearing on the forearms, which also improves proximal joint stability and enhances disassociation of the hand from the forearm, while the student is engaged in forming letters. The prone position is most stressful, demanding more cocontraction through the neck and upper-back musculature than either standing or sitting and tires many children quickly. Some students resist this position by sidelying or maintaining a quadruped position. The occupational therapist must be discerning of the child's tolerance to this position; lying prone for 5 minutes may be used as a benchmark. Both of these nontraditional writing positions elicit a variety of responses from children. They may range from marked enthusiasm to verbal opposition. Students with severe attention deficits may initially become more distractible and disorganized as well.

## Biomechanical Approach

Sitting posture, pencil grasp, and paper position are all ergonomic factors the occupational therapist must attend to with the student with handwriting dysfunction. Although standing and lying prone may be encouraged as alternative writing positions in the classroom, children spend much of their day seated at a desk and must be properly fitted in their desk. Benbow (1990) recommends the child be seated at a height so both feet are firmly planted on the floor, providing support for weight shifting and postural adjustments while writing. She also suggests the desk surface to be at a height 2 inches above the flexed elbow when the child is seated in the chair. This position allows the child stability and symmetry for performing written classwork (Benbow, 1990).

Pencil grasp and its relationship to functional writing remain an uncertainty for occupational therapists and educators. Studies indicate a variation of pencil grasps existing among nondysfunctional adults and children (Bergmann, 1990; Schneck, 1991; Schneck & Henderson, 1990; Ziviani, 1983; Ziviani & Elkins, 1984), despite the traditional advocacy of the dynamic tripod grasp for writing. For children with handwriting dysfunction, pencil grasp should be carefully observed. However, atypical grasp alone is not necessarily a precursor to poor writing (Ziviani). How much time and energy should be devoted to developing a mature pencil grasp (dynamic tripod or lateral tripod) remains questionable, but functional handwriting probably has more to do with the dynamic balance of the extrinsic and intrinsic muscles and precision control than a child's prehensile pattern (Boehme, 1988; Ziviani).

Many prosthetic devices exist for assisting the child to develop a mature pencil grasp. Triangular pencil grips, Stetro grips, My-Grip Personal grips, thick-barreled pencils, and orthoplast pencil holders (Weiser, 1986) are all designed to encourage a tripod grasp. For the child with a closed web space and/or typically flexed fingers, an adaptive grip device or wider-barreled pencil may assist in alleviating the muscle fatigue and tension when writing. Children exhibiting an incorrect grip with a stable opposed thumb may benefit from a Stetro grip to correct finger positions (Benbow, 1990; Meyers, this volume). Ziviani (1987) recommends that the focus of handwriting intervention be placed on the balance of finger movements and that these devices be transitory. Another position that facilitates more dynamic movement of the tripod grip fingers is holding a small ball or eraser with the ring and little fingers. This posture adds stability to the ulnar side, resulting in an increase of mobility of the radial side.

Conversely, the older student with hypotonic hand musculature may benefit from an adapted grasp, by placing the shaft of the pencil into the space between the middle and index fingers and supporting it with the pads of the thumb, index finger, and middle finger (Benbow, 1990) (see Figure 8). Ziviani (1987) suggests that if pencil grasp is being modified, the intervention should occur at a young age, because the longer the motoric pattern is present, the more difficult it is to remediate.

The position of the paper for the right-handed

Figure 8. An Adapted Grasp May Be Effective for the Child with Hypotonicity or Weak Hand Musculature

Figure 9. Paper Positions for (a) Right-Handed Writers; (b) Left-Handed Writers

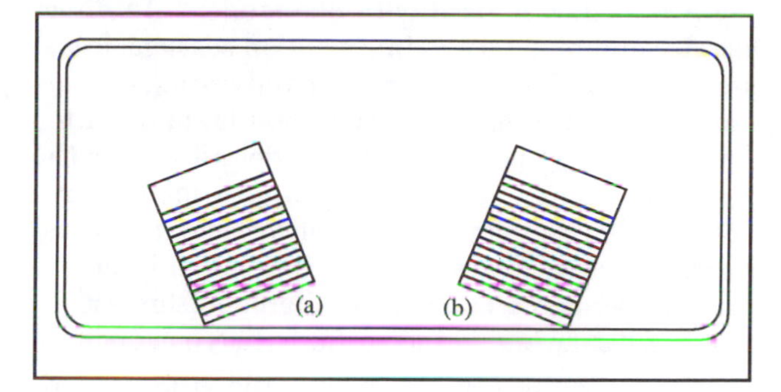


writer should be angled between 20 and 35 degrees from parallel to the edge of the desk with a slant to the left and the paper placed to the right of the midline. The mirror image is desirable for the left-handed writer. See Figure 9 for the recommended paper positions for right-handed and left-handed writers. The writing utensil should be held below the line, and the nonpreferred hand should stabilize the writing paper (Alston & Taylor, 1987). Left-handed writers may need to tilt their papers closer to 35 degrees, allowing for better visibility of their written product.

## Motivational Approach

Building a positive rapport with the student is key for any occupational therapy intervention program. Specific to handwriting, sharing with the child about the importance of handwriting and the reasons for the intervention, such as "we need to help your fingers catch up to your brain so you can learn how to write" (S. Woodrum, personal communication, April 26, 1991), may spur an initial sense of trust. Frequently, students become more motivated to exert a sincere effort when someone else is invested in this area of their lives.

The student's interest in writing may increase in an intervention program that is accentuated with varied and new activities, encouragement and reinforcement, and attainable expectations. Allowing the student success, choices, and responsibility are empowering and motivating forces leading toward more functional, legible handwriting.

## Comprehensive Handwriting Intervention Program

Developing and implementing effective and comprehensive handwriting intervention programs for students requires attention to all aspects of handwriting. Integrating the neuromuscular, acquisitional, multisensory, biomechanical, and motivational approaches challenges the occupational therapist's expertise, craft, and savvy in devising a creative and exciting intervention unique for each student with handwriting difficulties. Through deliberate organization and structure of the intervention, occupational therapists can:

1. Prepare students posturally for writing in the therapy room and the classroom,
2. Provide handwriting instruction in a sequential, organized format encompassing perceptual-motor development,
3. Vary the sensory input during handwriting instruction to include an assortment of mediums, materials, and positions for students,
4. Guide and direct the students to use mechanisms, approaches, and practical strategies for functional handwriting, and
5. Provide choices, encouragement, and reinforcement for each student engaged in the handwriting program.

## Delivering Occupational Therapy Services

The American Occupational Therapy Association (1989) has recognized service provision models (direct service, monitoring, and consultation) for the public school setting. Service provision, according to Public Law 94-142, is intended to be based solely on the student's needs as determined by the child's study team, and not on parental desires or school district resources, such as personnel or space (Dunn, 1988). Therefore, when considering the student's academic needs (e.g., improving handwriting), the occupational therapist will need to determine which service is most appropriate. Although the most familiar model of service provision is direct service, requiring direct interaction between the student and the occupational therapy practitioner to meet the student's educational need (AOTA, 1989), handwriting intervention may be effectively provided in the models of monitoring and consultation. Occasionally, a combination of service models is used with one student. In direct service and monitoring, intervention sessions focusing on postural preparation and handwriting instruction should be implemented on a regular and routine basis—daily, if possible. This model allows for the consistent, frequent practice essential to motor learning of such a complex task as handwriting. Individual or group sessions are possible. Individual intervention may be appropriate, but a small group of three or four children can be highly successful, satisfying, and fun!

Direct service might involve the student with hypertonia, severe motor planning problems, and very poor visual-perceptual skills who is demonstrating illegible handwriting. The therapist's role in intervention is critical to reduce the child's general muscle tone through slow rocking on the therapy ball and to release a tight scapula by pressing and gently rotating it until release is felt by the student and the student is posturally prepared to write. Along with the hypertonicity, the student's dyspraxia and poor visual perception influence handwriting. Writing instruction may need to be modified by focusing on the action needed to form letters and augmented by using a consistent set of visual and auditory cues. Thus, ongoing clinical judgments are needed to adjust the activities to best meet a student's ongoing needs and best performed by direct service from the occupational therapy practitioner.

Monitoring the teaching, training, and supervising of other persons who implement the intervention program (AOTA, 1989) is a viable means of delivering handwriting intervention to children. The occupational therapy practitioner will need to maintain regular contact (at least twice a month) with the service provider to determine if adjustments in intervention procedures are needed (AOTA). To ascertain whether handwriting intervention can best be delivered through this model, the therapist will need to consider: (1) the trainee's ability to demonstrate the procedures without assistance, (2) the trainee's capacity to communicate the risks, restrictions, and signals of failure warranting termination of the procedures and contact with the supervising therapist (AOTA; Dunn, 1988), and (3) the student's appropriateness for a program implemented by a person other than an occupational therapist. A child exhibiting low muscle tone, slight proximal joint stability, and poor visuomotor control and presenting poor handwriting may be a candidate for a monitored program. The therapist might train an educator, paraprofessional, or school volunteer to routinely implement the handwriting intervention program. Frequently, the service provider is able to administer the program more often than the occupational therapist.

Lastly, consultation is a service model in which the occupational therapist provides expertise to support the operation of the school system (AOTA, 1989). Related to handwriting, *case consultation*, which focuses on the student's educational needs, might involve designing proper classroom seating, providing an adaptive grip for a pencil, and suggesting an approach for better horizontal alignment of letter forms (e.g., raised lines on paper). In *colleague consultation*, a therapist might improve the skills and knowledge of all 1st-grade teachers by providing an inservice program regarding handwriting, the readiness skills needed, and alternative materials and media for instruction. A contribution to the school district via *system consultation* may involve the occupational therapist serving on a handwriting curriculum review committee for elementary education.

## Summary

In the educational setting, students with learning disabilities, developmental delays, and neurological impairments often struggle to write legibly. Frequently, these children are referred to occupational therapy for evaluation and possible intervention. When administering a comprehensive assessment, therapists must consider the underlying sensorimotor foundations of handwriting; the domains, legibility components, and ergonomic factors of the child's handwriting; and the interaction of psychosocial and cognitive behaviors of the student. Once these areas are determined, an integrated handwriting intervention program may be developed by the occupational therapist and the education team. Intervention programs characterized by postural preparation of the student for writing, sequenced handwriting instruction with a variety of writing materials and media, practical biomechanical strategies for writing, and encouragement and reinforcement from the service provider can be successful, rewarding, and fun for both the student and the occupational therapy practitioner.

## Acknowledgments

I would like to thank Beverly Ingram, OTR/L, my Alaskan colleague, and the children of the Kenai Peninsula Borough School District in Homer, Alaska for sparking and encouraging an interest in one of children's most important academic occupations—handwriting.

## References

Alston, J. (1983). A legibility index: Can handwriting be measured? *Educational Review, 35,* 237–242.

Alston, J., & Taylor, J. (1987). *Handwriting: Theory, research and practice.* London: Croom Helm.

American Occupational Therapy Association. (1989). *Guidelines for occupational therapy services in school systems.* Rockville, MD: Author.

Anderson, P.L. (1983). *Denver handwriting analysis.* Novato, CA: Academic Therapy Publications.

Armitage, D., & Ratzlaff, H. (1985). The non-correlation of printing and writing skills. *Journal of Educational Research, 78,* 174-177.

Barbe, W.B., Milone, M.J., & Wasylyk, T. (1983). Manuscript is the write start. *Academic Therapy, 18,* 397-405.

Benbow, M. (1990). *Loops and other groups.* Tucson, AZ: Therapy Skill Builders.

Bergman, K.E., & McLaughlin, T.F. (1988). Remediating handwriting difficulties with learning disabled students: A review. *B.C. Journal of Special Education, 12*(2), 101-120.

Bergmann, K.P. (1990). Incidence of atypical pencil grasps among nondysfunctional adults. *American Journal of Occupational Therapy, 44*(8), 736-740.

Boehme, R. (1988). *Improving upper body control.* Tucson, AZ: Therapy Skill Builders.

Cook, D.G. (1991). The assessment process. In W. Dunn (Ed.), *Pediatric occupational therapy: Facilitating effective service provision* (pp. 35-72). Thorofare, NJ: Slack.

Dunn, W. (1988). Models of occupational therapy service provision in the school system. *American Journal of Occupational Therapy, 42*(11), 718-722.

Dunn, W., & Lane, S. (1986, February). *A neurobiological foundation for sensory integration.* Course conducted at Stanford Children's Hospital, Palo Alto, CA.

Duvall, B. (1985). *Evaluating the difficulty of cursive. manuscript, italic and D'Nealian handwriting* (Report No. CS 209 484). (ERIC Document Reproduction Service No. ED 265 539).

*Education of All Handicapped Children Act of 1975* (Public Law 94-142), 20 U.S.C. 1401.

Erhardt, R.P. (1982). *Developmental hand dysfunction.* Laurel, MD: RAMSCO Publishing Company.

Exner, C.E. (1989). Development of hand functions. In P.N. Pratt & A.S. Allen (Eds.), *Occupational therapy for children* (pp. 235-259). St. Louis, MO: Mosby.

Getman, G.N. (1985). Hand-eye coordination. *Academic Therapy, 20*(3), 261274.

Graham, S., & Miller, L. (1980). Handwriting research and practice: A unified approach. *Focus on Exceptional Children, 13,* 1-16.

Hagin, R.A. (1983). Write right—or left: A practical approach to handwriting. *Journal of Learning Disabilities, 16*(5), 266-271.

Hammill, D.D. (1986). Correcting handwriting deficiencies. In D.D. Hammill & N.R. Bartel (Eds.), *Teaching students with learning and behavior problems* (pp. 155-177). Newton, MA: Allyn & Bacon.

Holm, M. (1986). Frames of reference: Guides for action—occupational therapist. In H.S. Powell (Ed.), *PILOT: Project for Independent Living in Occupational Therapy* (pp. 69-78). Rockville, MD: American Occupational Therapy Association.

King-Thomas, L., & Hacker, B.J. (1987). *A therapist's guide to pediatric assessment.* Boston, MA: Little, Brown.

Kirk, S.A., & Chalfant, J.C. (1984). *Academic and developmental learning disabilities.* Boston: Houghton Mifflin.

Lamme, L.L., & Ayris, B.M. (1983). Is the handwriting of beginning writers influenced by writing tools? *Journal of Research and Development in Education, 17*(1), 33-38.

Larsen, S.C., & Hammill, D.D. (1989). *Test of legible handwriting.* Austin, TX: Pro-Ed.

Laszlo, J.I., & Bairstow, P.J. (1984). Handwriting: Difficulties and possible solutions. *School Psychology International, 5,* 207-213.

Lindsay, G.A., & McLennan, D. (1983). Lined paper: Its effects on the legibility and creativity of young children's writing. *British Journal of Educational Psychology, 53,* 364-368.

Maddox, V. (1986). Postural preparation for writing. *American Occupational Therapy Association Developmental Disabilities Special Interest Section Newsletter, 9*(3), 3-7.

McGourty, L.K., Foto, M., Marvin, J.K., Smith, N.M., Smith, R.O., Kronsnoble, S., & Strickland, L.R. (1989). *Uniform terminology for occupational therapy-second edition.* Rockville, MD: American Occupational Therapy Association.

Milone, M.N., Jr., & Wasylyk, T.M. (1981). Handwriting in special education. *Teaching Exceptional Children, 14(2),* 58-61.

Oetter, P. (1986, July). *Camp Avanti for children with sensory integrative dysfunction.* Santa Barbara, CA.

Olsen, J.Z. (1980). *Handwriting without tears.* Brookfield, IL: Fred Sammons.

Pasternicki, J.G. (1987). Paper for writing: research and recommendations. In J. Alston & J. Taylor (Eds.), *Handwriting: Theory, research and practice* (pp. 68-80). London: Croom Helm.

Peck, M., Askov, E.N., & Fairchild, S.H. (1980). Another decade of research in handwriting: Progress and prospect in the 1970s. *Journal of Educational Research, 73*(5), 283-298.

Phelps, J., & Stempel, L. (1987). *The children's handwriting evaluation scale for manuscript writing.* Dallas, TX: Texas Scottish Rite Hospital For Crippled Children.

Phelps, J., Stempel, L., & Speck, G. (1984). *The children's handwriting evaluation scale: A new diagnostic tool.* Dallas, TX: Texas Scottish Rite Hospital for Crippled Children.

Price, A. (1986). Applying sensory integration to handwriting problems. *American Occupational Therapy Association Developmental Disabilities Special Interest Section Newsletter, 9*(3), 4-5.

Rosenbloom, L., & Horton, M.E. (1971). The maturation of fine prehension in young children. *Developmental Medicine and Child Neurology, 13,* 3-8.

Rubin, N., & Henderson, S.E. (1982). Two sides of the same coin: Variations in teaching methods and failure to learn to write. *Special Education: Forward Trends, 9*(4), 17-24.

Schneck, C.M. (1991). Comparison of pencil-grip patterns in first graders with good and poor writing skills. *American Journal of Occupational Therapy, 45*(8), 701–706.

Schneck, C.M., & Henderson, A. (1990). Descriptive analysis of the developmental progression of grip position for pencil and crayon control in nondysfunctional children. *American Journal of Occupational Therapy, 44*(10), 893-900.

Stott, D.H., Moyes, F.A., & Henderson, S.E. (1985). *Diagnosis and remediation of handwriting problems.* Guelph, Ontario: Brook Educational Publishing.

Taylor, J. (1985). The sequence and structure of handwriting competence: Where are the breakdown points in the mastery of handwriting? *British Journal of Occupational Therapy, 48*(7), 205-207.

Thurber, D. (1983). *D'Nealian manuscript—An aide to reading development.* (Report No. CS 007 057). Grand Rapids, MI: Paper presented at Annual Meeting of the Michigan Reading Association. (ERIC Document Reproduction Service No. ED 227 474).

Weiser, D. (1986). Handwriting: Assessment and treatment. *American Occupational Therapy Association Developmental Disabilities Special Interest Section Newsletter, 9*(3), 1-3.

Ziviani, J. (1983). Qualitative changes in dynamic tripod grip between seven and 14 years of age. *Developmental Medicine & Child Neurology, 25,* 778-782.

Ziviani, J. (1987). Pencil grasp and manipulation. In J. Alston & J. Taylor (Eds.), *Handwriting: Theory, research and practice* (pp. 24-39). London: Croom Helm.

Ziviani, J., & Elkins, J. (1984). An evaluation of handwriting performance. *Educational Review, 36*(3), 241-269.

# 6

# Developing Scissors Skills in Young Children

*Colleen Schneck*

*Carmela Battaglia*

## Importance of the Use of Scissors

"Tool use is a purposeful, goal-directed form of complex object manipulation that involves the manipulation of the tool to change the position, condition, or action of another object" (Connolly & Dalgleish, 1989, p. 895). Tools enhance the proficiency with which skills are performed and are needed for feeding, writing, grooming, and a variety of other tasks basic to activities of daily living. Mastery of tool use is an important developmental skill that affects a child's functional performance. Without this ability to successfully manipulate tools, the child may be unable to perform many daily tasks without assistance.

One such tool is scissors. Mastering the use of scissors is a necessary component of fine-motor skill development in children (Stephens & Pratt, 1989), which occurs during the preschool years. Scissors skills are needed for many functional activities of daily living and are required for many school-related tasks, including arts and crafts activities. Preschool, kindergarten, and early elementary teachers are concerned with the child's ability to perform fine-motor skills, including cutting with scissors, and devote a great deal of time to developing this skill. This skill development is often delayed in children with physical, emotional, and/or cognitive deficits. Children who demonstrate difficulties with fine-motor skills, which include the ability to use scissors, are often referred to occupational therapy. Occupational therapists possess the ability to assess the many factors involved in scissors skills, as well as the child's overall fine-motor and visuomotor skills. These factors involved in scissors skills include visual, tactile, and proprioceptive processing and their effects on hand function, motor planning, bilateral integration, postural adjustments, grip, strength, and skill acquisition. To accomplish this evaluation, it is important for the clinician to have knowledge of the normal development of scissors skills.

This chapter familiarizes the clinician with the normal developmental sequence of scissors skills in children, including prerequisite skills, recalibration, and the motoric aspects of scissors skills. Evaluation tools available for assessing these skills will be described and critiqued, and clinical evaluation of the sensorimotor aspects of cutting with scissors will be included. Therapy for the underlying sensorimotor problems that result in difficulties using scissors deficits will be discussed, as well as types of scissors and practice books available. Recommendations for further research in scissors-skill development, evaluation, and treatment will be suggested.

## Developmental Sequence

### Prerequisite Skills

Cutting with scissors demands proficiency in eye-hand coordination, motor planning, bilateral coordination, hand and finger dexterity, and tool use. In a nonhandicapped child, these scissors skills develop in a typical sequence that may be influenced by maturational rate and environmental stimulation.

Klein (1987) describes eight prerequisite skills that are essential for competent use of scissors. The child must have reached the constructive developmental play stage that allows him or her to attend, cooperate, and interact with toys, including a demonstrated interest in scissors. The ability to open and close the hand, as well as range of motion of the fingers is important for holding and manipulating scissors. Bilateral hand use is important as the preferred hand must hold and coordinate the scissors while the other hand holds, stabilizes, and rotates the cutting material. Eye, hand, and arm coordination with proximal stability of the shoulder, elbow, and wrist are essential for smooth coordinated movements. Sensorimotor feedback of kinesthetic, tactile, vestibular, and visual information is needed for motor planning and postural adjustments when cutting with scissors. A very important component of sensorimotor feedback is calibration.

Calibration is a sense of weight and distance of an arm or leg in space. It involves the ability to coordinate muscular effort to move a limb in space while receiving somatosensory and visual feedback (Schwartz & Reilly, 1981). Without a well-developed sense of calibration, one would have difficulty localizing the hand in space during fine-motor activity. Recalibration is "the use of somatosensory and visual information to modulate and adjust muscular effort to compensate for the added weight and length of tools" (Schwartz & Reilly, p. 16). For example, when scissors are introduced, recalibration occurs to allow for the added weight and length of this tool to the limb. Without recalibration, the skill and performance of the task would be compromised.

Recalibration continues to develop as the child matures and develops his or her coordination. It is a developmental process occurring between the development of manipulative prehension and the acquisition of skilled use of a tool. Recalibration using shorter/lighter-weight tools precedes recalibration of larger/heavier-weight tools; therefore, the use of small scissors is recommended before introducing large scissors (Schwartz & Reilly, 1981).

Tool skill also develops sequentially (Schwartz & Reilly, 1981) beginning with the development of grasp and release of objects. Next, the individual develops calibration of his or her upper extremities in space. When a tool is introduced, recalibration of the limb in conjunction with the tool begins to develop. Lastly, the child develops skill in manipulating the tool's working edge. Guiding the working edge is usually a problem for children who demonstrate an inappropriate or weak scissors grasp. When grasp is weak, a child often holds the scissors at an angle that interferes with its cutting edge. There are several types of scissors commercially available, many of which will continue to cut even if the scissors are held at an incorrect angle. Various types of scissors and their characteristics are described in the treatment section.

## Development of Scissors Skill

A skill is defined as an ability that demonstrates dexterity and is goal directed (Connolly & Dalgleish, 1989). Through the mechanism of sensorimotor feedback, provided through practice, a skill progresses from a cognitive task to an automatic activity. What is important is that, without practice, a fine-motor task such as cutting with scissors, will not develop to the level of skilled performance necessary for daily life tasks.

"Skill" includes a wide range of sensorimotor performances such as perceptual motor, fine dexterous movements, and problem solving (Connolly, 1973). These sensorimotor performances, which are organized into a sequence of goal-directed activities that are either guided by or corrected by feedback, are considered skilled behaviors (Fishbein, 1976). Almost all skilled movements involve postural, transport, and manipulation components (Smith & Sussman, 1969). Scissors use involves the maintenance of body posture in space, movement of the arm to transport the hand across the page, and use of the fingers to manipulate the scissors in order to cut the paper.

The nature of skill acquisition can be divided into three specific characteristics (Welford, 1968): (1) building and organizing coordinated activity, (2) modifying the data needed to be fed into the system via a gradual learning process, and (3) combining anticipatory behavior and dynamic interplay between receptor-effector functions. Coley (1978) has applied this skill process to learning how to use scissors. First the child must plan the positioning of his or her hands on the scissors and paper. While cutting, the child receives feedback from the tactile, kinesthetic, and visual systems. When first learning this skill, the child has a conscious awareness of the movement, he or she monitors the progress of the plan through the feedback provided, and modifies and refines the motor plan as needed.

The development of tool use is important in skill development. The initial period of learning a skill involves assuming the correct posture and holding the tool in its optimal position for function (Connolly, 1973). The development of scissors grip will be discussed in the motoric aspects section. In addition to postural control and correct tool position, scissors skills require coordination of the tactile, kinesthetic, visual, and motor systems. Continued improvement in scissor skills occurs with maturation as well as with experience, and tends to follow an orderly developmental sequence. Therefore, the actual attainment of scissors skill is the result of a combination of maturation and practice and thus this complex skill develops slowly over time.

Scissors skill has been described as a series of progressive steps that are usually mastered in a sequential order (Klein, 1987; Stephens & Pratt, 1989). A child begins by showing an interest in using scissors. Next he or she learns to hold, open, and close them. His or her first attempts at cutting are generally short, random snips along the edge of a piece of paper while holding the paper in the nonpreferred hand. He or she then progresses to cutting in a forward direction that requires more than one snipping motion.

Once he or she can cut a straight line, he or she will attempt curves, zig-zag lines, and adapt cutting movements to accommodate for varying line lengths within the same cutting task. Then he or she will learn to cut geometric shapes, simple figures, and, finally, complex figures. Once the child becomes proficient in cutting paper with scissors, he or she progresses to cutting other nonpaper materials such as cloth or tape.

Descriptions of the developmental sequence for the mastery of specific scissors-cutting skills varies from author to author (Bleck, 1975; Quick & Campbell, 1983; Stephens & Pratt, 1989). The skills described vary in both the age level by which they should be achieved and the sequential order in which the skills are accomplished. Most authors agree that a child first learns to hold scissors, then progresses to snipping, and then to cutting in a straight line. Variation reported between authors occurs in the order in which he or she learns to cut shapes, how close to the line he or she must be, and how much time should be allowed for each task, the last two being measures of skill.

## Gender Differences

Schwartz and Reilly (1981) suggest that most children do not cut well until 6 years of age, with girls developing earlier than boys due to maturational differences and home practice. Karr (1934) found that children of the same age who cut at home, usually girls, were able to cut more difficult shapes than those who had little practice at home. Karr also found that when matched by 6-month-age intervals for gender, the girls were slightly higher in cutting scores but the difference was not significant. Campbell and Early (1980) found no significant difference in cutting ability between boys and girls; however, sex differences were noted in all age categories except the 6-year-olds. In addition, Lopez (1986) found no significant differences between boys and girls in the motoric components of cutting with scissors.

## Motoric Components

An important aspect in the development of scissors skills in which occupational therapists are concerned is the motor component. This includes prehension patterns, arm position, bilateral coordination, strength, associated reactions, and postural adjustments. Unfortunately, research has been limited to identifying some developmental milestones for certain aspects of cutting and little information is available on the motoric aspects (Lopez, 1986).

Holle (1981) has described the manipulation of scissors as a difficult task, requiring a high level of motor ability, eye-hand coordination, and finger strength. In addition, the use of scissors has been described as a complex bilateral skill (Petrone, 1976). There is some agreement on what the appropriate scissors grip should be (Fairbanks & Robinson, 1967; Holle; Klein, 1987; Larson, 1976; Marshall, 1975; Myers, this volume). The mature grip presupposes the ability to isolate thumb and fingers. The thumb of the preferred hand is inserted in one loop, while the middle, ring, or middle and ring fingers (depending on the size of the scissors' loops) are inserted in the other. The index finger is placed below the bottom loop to provide stability and strength and to direct the cutting activity.

One study (Lopez, 1986) was conducted to measure some aspects of the motoric components of the skill of cutting with scissors in children aged 2 to 6 years. The subjects were 60 normal, right-handed children with 6 boys and 6 girls in each age group. The Cutting with Scissors Movement Rating Scale (CSMRS), which consisted of five subsections, was used. These five subsections were: (1) isolated thumb movement at the proximal joint, (2) opening and closing of the scissors, (3) positioning of the scissors during cutting, 4) upper-extremity control of the arm manipulating the scissors, and (5) bilateral coordination.

The results of this study provide valuable descriptive information on these motoric components of scissors use. Overall, the most rapid development of postural and movement components of performance occurred between 2 and 3 years of age. There seemed to be more variability in the development of the motoric aspects in younger children. No significant differences were found between the performance of boys and girls on any of the five subtests. From the 2.6- to 2.11-year age range on, all of the children were able to isolate thumb movements. The skill of opening and closing scissors without paper present appears to develop from the child mainly using one hand with assistance from the other, to the child using one hand, opening the scissors less than halfway, in a smooth fashion. A middle step in this development shown by 8 of the 60 children consisted of the child using one hand, opening the scissors halfway or more.

The positioning of the scissors during the cutting section of the CSMRS reveals much needed information on the development of grip used for cutting. Table 1 provides descriptive information on these grips in developmental order. Figure 1 illustrates the grips described in Table 1. Younger children showed a

**Table 1. Number of Children in Age Groups (n = 6) Who Demonstrated Developmental Stages of Scissors Grip***

| | Age Groups | | | | | | | | | |
|---|---|---|---|---|---|---|---|---|---|---|
| Grip on scissors | 2.0–2.5 | 2.6–2.11 | 3.0–3.5 | 3.6–3.11 | 4.0–4.5 | 4.6–4.11 | 5.0–5.5 | 5.6–5.11 | 6.0–6.5 | 6.6–6.11 |
| Thumb, one loop; index, other; rest of fingers extended | 3/6 | 3/6 | 1/6 | | | | | | | |
| Thumb, one loop; index and middle in other; rest of fingers extended | 3/6 | 1/6 | | | | | | | | |
| Thumb, one loop; middle, other; rest of fingers flexed; index not stabilizing scissors | | | 2/6 | | | | | | | |
| Thumb, one loop; index, other, rest of fingers flexed | 3/6 | 1/6 | 2/6 | 3/6 | 1/6 | 5/6 | 2/6 | 2/6 | 1/6 | |
| Thumb, one loop; index and middle in other; rest of fingers flexed | | 1/6 | 1/6 | 3/6 | 5/6 | 1/6 | 4/6 | 4/6 | 4/6 | 4/6 |
| Thumb, one loop; middle other; index stabilizing | | | | | | | | | 1/6 | 2/6 |

*Lopez, M. (1986). *Developmental sequence of the skill of cutting with scissors in normal children 2 to 6 years old. Unpublished Master's thesis, Boston University.*

**Figure 1. Hierarchy of Developmental Scissors Grips***

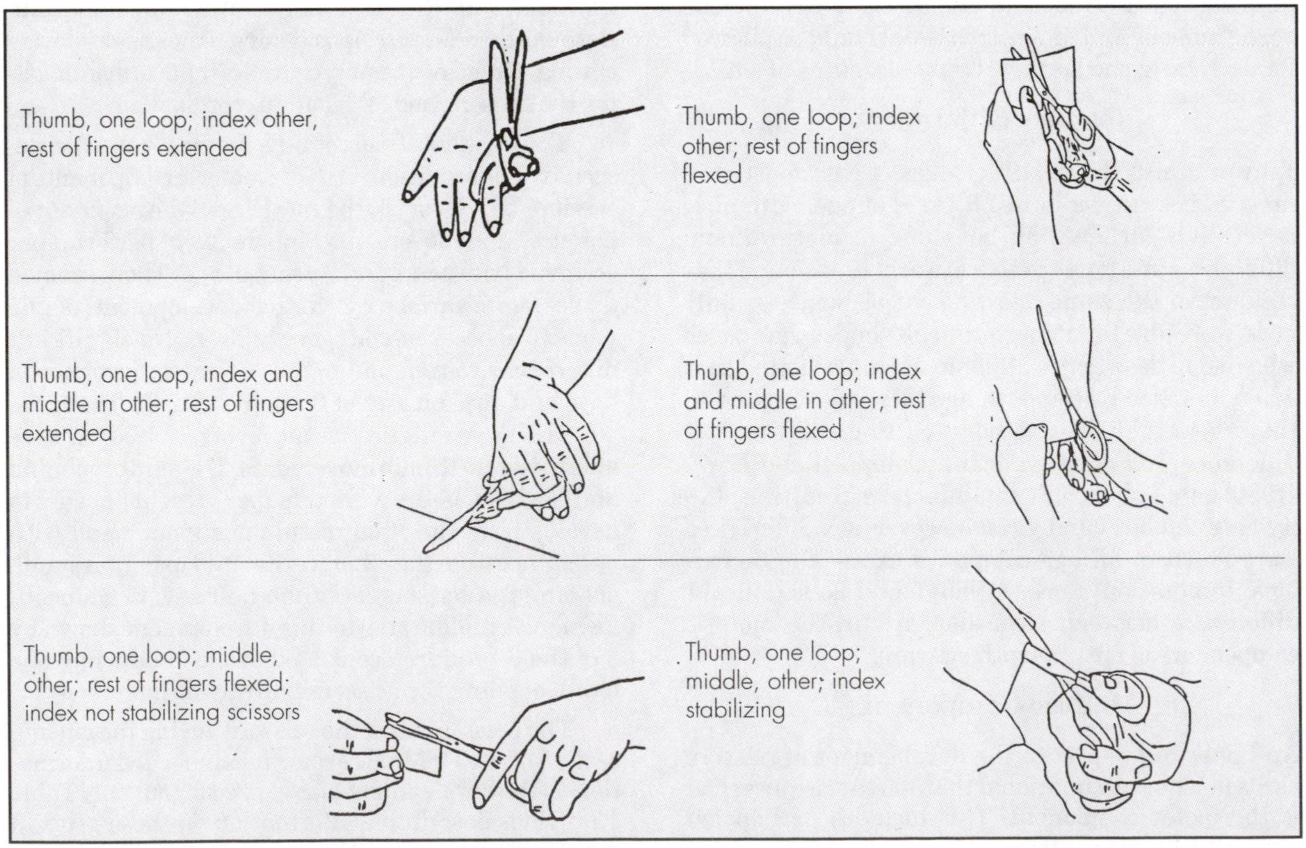

Thumb, one loop; index other, rest of fingers extended

Thumb, one loop; index other; rest of fingers flexed

Thumb, one loop, index and middle in other; rest of fingers extended

Thumb, one loop; index and middle in other; rest of fingers flexed

Thumb, one loop; middle, other; rest of fingers flexed; index not stabilizing scissors

Thumb, one loop; middle, other; index stabilizing

*As described by Lopez (1986).

higher variability of grip patterns used. There was also a progression of extension to flexion of the free fingers, with the free fingers progressed from posturing in extension to flexion, with extension in all the 2.0- to 2.5-year-old children. Finger extension decreased with age and ceased by 3.6 to 3.11 years. Only 3 of the 12 children in the 6-year-old age groups showed a mature grip as defined in the literature.

The upper-extremity control of the arm manipulating the scissors subtest showed a clear developmental progression of the arm position for manipulating the scissors. Younger children position the forearm in pronation, changing to mid-position by 3.6 to 3.11 years. An interesting observation is that all children achieved the highest score on arm position, yet very few children in the 6-year-old age groups had a mature scissors grip. The author suggests that a long time elapses between proximal and distal control of the arm in manipulating scissors. Another possible explanation for the separate development of arm position and hand grip may be that proximal and distal motor control are operated by different central nervous system mechanisms (see Pehoski, this volume).

A clear, developmental progression is also noted with bilateral coordination. This progression begins with the child stabilizing the paper: (l) in an inappropriate static fashion (i.e., fingers in path of scissors), (2) in an appropriate static fashion, (3) in a dynamic fashion but without rotating the paper, to (4) a dynamic fashion, including rotating the paper. By ages 5.6 to 5.11 years, all but one child showed the most mature pattern.

This study by Lopez (1986) has provided valuable descriptive information on what motor patterns children use to cut with scissors. Due to the limited number of subjects, however, further study is needed on these motoric aspects. Research with a large sample would help establish norms. In addition, expanding the age range would help determine when the mature scissors grip develops. Variations of these grips have been noted by Schneck (in progress) in studying scissors grip patterns in learning-disabled children. Figure 2 provides descriptive information on some of the grips observed. However, it is not known if these are normal variations of scissors grips. Therefore, further information is needed on the normal variations of scissors grips in order to better evaluate scissors grips in children with fine-motor problems.

A study was conducted by Sellers (1983) to determine whether a relationship existed between hand strength and selected pencil-scissors skills in kindergarten-aged children. Three fine-motor tasks were selected, which included cutting, tracing, and drawing from the Bruininks-Oseretsky Motor Proficiency Test (Bruininks, 1978). Following the above tasks, the child's grip and lateral pinch were measured with

**Figure 2. Grip on Scissors Noted in Learning-Disabled Students Aged 6 to 12 Years**

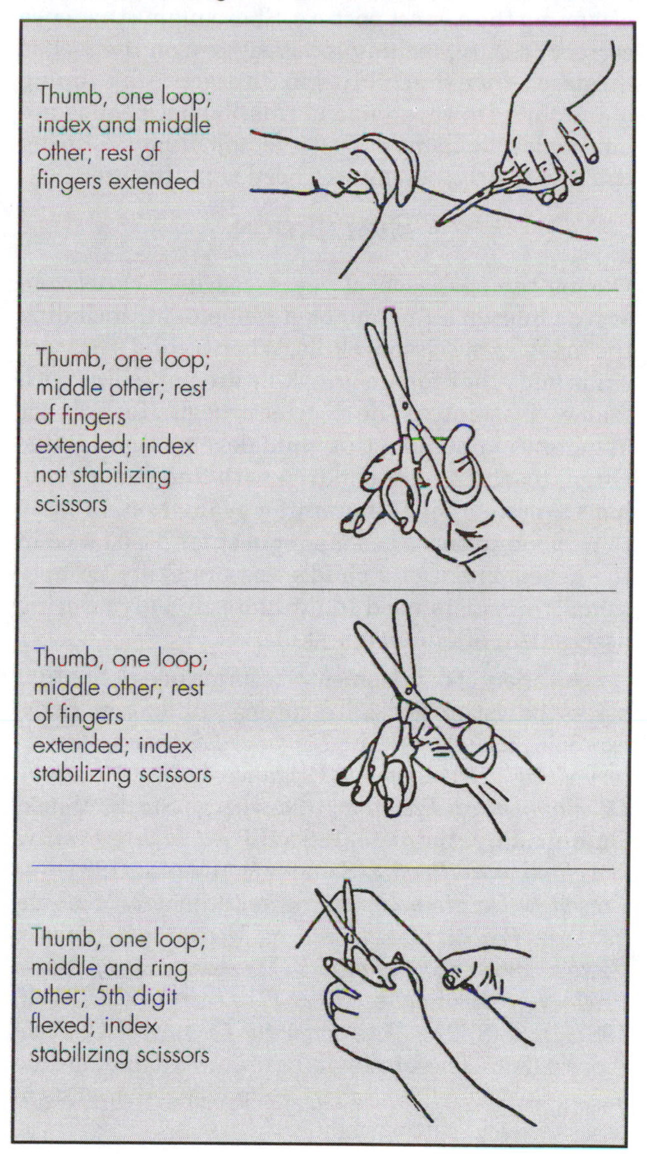


a hydraulic hand dynamometer. The two strength measurements were significantly related. However, there were no significant relationships between the strength measurements and pencil-scissors tasks, indicating that the overall strength of the hand was not significantly related to the child's ability to use pencils and scissors to produce well-controlled hand movements.

Associated reactions during cutting with scissors have been reported in normal 2-year-old children (Stutsman, 1948), including similar movements of the opposite hand, an open mouth, and the tongue protruding. Lopez (1986) reported associated reactions in the arm holding the paper and around the mouth in younger age groups; however, no systematic record was kept of these observations and, therefore, further analysis could not be done.

Connolly (1973) has described that in the initial period of learning a skill the child is concerned with assuming the correct posture. She explains that once correct posture, including correct grip on the tool, is mastered, then the child begins to practice fine tuning of the skill. However, no information has been found on the development of postural adjustments during cutting; clearly, research is needed in this area.

## Evaluation

During the early school years, teachers closely observe children's fine-motor development, including the mastery of scissors skills. When a child's scissors skills and other fine-motor skills are not in line with fellow classmates, the teachers begin to look for assistance in helping this child develop these skills. Often, teachers refer children with fine-motor problems to occupational therapy for evaluation. In order to respond to the teacher's request for assistance in the development of a child's scissors skills, occupational therapists need to evaluate this area during assessment of fine-motor skills.

A variety of developmental tests includes items that assess the visuomotor skill of cutting with scissors. Some examples include: (1) The *Brigance Diagnostic Inventory of Early Development* (Brigance, 1978); (2) *Battelle Developmental Inventory* (Newberg, Stock, Wnek, Guidubaldi, & Suinicki, 1984); (3) *Early Intervention Developmental Profile* (Schafer & Moersch, 1981); (4) *Preschool Developmental Profile* (Schafer & Moersch, 1981); (5) *Peabody Developmental Motor Scales* (Folio & Fewell, 1983); (6) *Bruininks-Oseretsky Test of Motor Proficiency* (Bruininks, 1978); (7) *Learn to Cut* (Wolf, 1987); and (8) *The Developmental Cutting Task Scale* (Lopez, 1986). These tests will be reviewed as they are the ones most frequently used by occupational therapists to assess scissors-cutting abilities.

The *Brigance Diagnostic Inventory of Early Development* (Brigance, 1978) is a criterion-referenced test of school-readiness abilities in children aged 2 to 7 years. The cutting-with-scissors subtest has 13 items with a minimum of 2 items at each age level. The directions are clear and understandable and the subtest is discontinued after two consecutive failures. This is one of the most useful assessments currently available for assessing scissors skills because a developmental age range for scissors skills can be determined. The authors note, however, that their research did not yield enough norming data to fully validate the developmental ages. These developmental ages are based on information from a developmental checklist presented in Bleck (1975), and responses from these authors' field tests.

The *Battelle Developmental Inventory* (Newberg et al., 1984), although a well-accepted assessment, contains only one item on scissors cutting, which is at the age of 3 to 4 years. Therefore, this assessment has limited usefulness in the assessment of scissors skills.

The *Early Intervention and the Preschool Developmental Profiles* (Schafer & Moersch, 1981) are both criterion-referenced tests that contain a perceptual/fine-motor section. For both of these assessments, the age range placement was based on the average of age norms of items that appeared on two or more standardized scales. On the *Early Intervention Developmental Profile*, there is one item at the 32- to 35-month age range, in which the child passes if the child makes a cut with scissors. On the *Preschool Developmental Profile*, there are three measures of scissors cutting, one at each of the following age levels: 3 1/2, 4, and 5 1/2 years. These items are: isolates thumb movements, cuts within 1/4 inch of straight and curved lines, and cuts out a picture of a whale within 1/4 inch.

The *Peabody Developmental Motor Scales* test (Folio & Fewell, 1983) is designed to measure gross- and fine-motor skills in infants and children with motor delays. This test evaluates skills the child has mastered, skills that are emerging, and skills that are not yet in the child's repertoire. This standardized assessment has good psychometric properties. There are five items assessing scissors skills in the eye-hand category. These include: 18 to 23 months—snipping with scissors; 30 to 35 months—cutting paper; 36 to 41 months—cutting lines; 42 to 47 months—cutting a circle; and 48 to 57 months—cutting a square. Although these items cannot be separated from the rest of the test in scoring, the therapist has an estimate of the age range of the child's scissors skills.

The *Bruininks-Oseretsky Test of Motor Proficiency* (Bruininks, 1978) is a well-standardized test of motor skills in children 4 1/2 to 14 1/2 years of age. On the visuomotor control subtest, there is an item assessing the child's ability to cut out a bold, outlined circle embedded within six consecutive circles. This score is incorporated into a composite fine-motor-skills score and there are no separate norms for the cutting item. Therefore, very little information is obtained on the child's scissors-cutting abilities except those observations made by the therapist.

Although *Learn to Cut* (Wolfe, 1987) is a structured program designed to teach the skill of cutting, pre- and posttest measures are available. In this program, the fine-motor task of cutting is divided into eight major skill areas, which include: snipping; cutting a line; cutting a simple shape—1/4-inch line; cutting a simple shape—1/8-inch line; cutting combined simple shapes—1/8-inch line; cutting circles and complex shapes—1/4-inch line; cutting circles and complex shapes—1/8-inch line; and cutting combined circles and complex shapes—1/8-inch line. This information is valuable if the therapist is interested in knowing where the child's skills are and how to progress in therapy, if a developmental age is not necessary.

The *Developmental Cutting Task Scale* (Lopez, 1986) was developed to obtain information on the perceptual-motor performance of cutting with scissors as measured by the precision of a cut on a line in a series of cutting tasks by children aged 2 to 6.11 years. The 23 items were developed based on a review of the literature. Due to the small number of subjects, norms could not be developed.

All of the above assessment tools only evaluate selected aspects of the visuomotor component of cutting with scissors. A more comprehensive assessment is necessary in order to evaluate these tools' use more accurately across the developmental sequence. This would include cutting accuracy and skill complexity in order to better determine treatment goals. In addition to evaluating the visuomotor aspect of cutting with scissors, clinical observations of the sensorimotor factors of this skill should be included in the assessment.

### Clinical Assessment of Scissors Skills

For the acquisition of scissors skills, a variety of sensorimotor factors as previously discussed must be considered. Limitations imposed by any one of these contributing factors, at any time in development, may affect the level of performance attained (Connolly, 1973). Therefore, these many factors should be considered when conducting a clinical assessment of scissors skills. Visual, tactile, and proprioceptive processing should be evaluated in terms of their effects on hand function. Children with decreased tactile and proprioceptive feedback from their hands continue to need constant visual monitoring in order to perform even simple straight-line cutting. Motor planning, including bilateral integration, and postural adjustments, as well as the child's recalibration abilities, should also be assessed.

Grip, arm position, and bilateral coordination can best be assessed using the criteria developed by Lopez (1986) until further research can be done to establish normative guidelines. Grips other than those observed by Lopez should also be noted and observed as all the normal variations of grips are not yet known. Associated reactions of the opposite hand and mouth should be noted if they occur in older children as an indicator of the amount of effort exerted by the child in performing the task.

The child's handedness should be determined prior to the cutting task in order to provide the child with appropriate right- or left-handed scissors. If a child does not have a preferred hand, then scissors could be provided that would cut no matter which hand was used. Handedness should be determined by means other than scissors use, since the use of scissors has been found to be an unreliable indicator of hand dominance (Kashihara, 1981; McFarland & Anderson, 1980). One measure that could be used to determine handedness is the *Lateral Consistency Test* adapted from Lyle (1976). The test consists of 10 items requiring the use of one hand (i.e., pointing to a spot on the wall). Lederer (1939) derived an index of handedness that indicates that a child with 80% hand preference significantly uses that hand more often than the other. Thus, if the child uses the same hand to perform 8 of the 10 items on the *Lateral Consistency Test*, it can be assumed that it is his or her preferred hand; otherwise, it can be assumed that the child does not have a preferred hand.

Although a significant correlation was not found between either hand strength or lateral pinch and pencil-scissors skills by Sellers (1983), grip strength should be evaluated as the results of this study may not apply to children with atypical muscle tone. Too forceful a grip may also interfere with scissors skills.

## Treatment Activities

A successful treatment plan to improve cutting with scissors should follow a developmental approach. One must take into consideration the child's developmental age, functional performance on a scissors evaluation, sensorimotor status, and chronological age so that activities are age appropriate, meaningful, and well-received by the child.

For those children who are not developmentally ready to begin cutting with scissors, precutting activities to develop hand strength, eye-hand coordination, and fine-motor dexterity are warranted. Playdough and theraplast activities will develop needed strength for cutting against resistance. Activities that require finger flexion, extension, and isolated movements will help prepare a child for opening and closing of scissors. Squirt guns, squeeze toys, and punching holes with a paper punch will help develop finger dexterity and strength. Cut-and-paste activities using bright and/or shiny paper may provide motivation for the child to want to cut.

Cutting with scissors is a bilateral hand activity requiring coordination and integration of the upper extremities. Bilateral activities should be encouraged prior to introducing scissors to the child. These may include using lacing or sewing cards, stringing beads, tearing paper, or snapping beads together. When cutting designs and shapes, bilateral coordination becomes important as the child is required to hold and turn the paper with the nonpreferred hand while the preferred hand guides the scissors in cutting. This is a skill that does not develop automatically and usually is mastered after considerable practice. Through demonstration (how to move the hand and rotate the paper), verbal cues, and physical hand-over-hand assistance, the child can be allowed to experience and develop the motor planning movements necessary to perform the cutting task.

## Scissors

As previously stated, the initial period of learning a skill is concerned with assuming the correct posture and holding the implement in its optimal position of function. Initially, the therapist should be concerned with selecting the appropriate scissors for the child in order to facilitate correct holding of the scissors. Once the child is posturally secure and tool position is mastered, exercises are used to practice and perfect the skill. External support may be needed when postural stability is not adequate (e.g., a chair with a high back and arm rests).

Selecting appropriate scissors is very important, as good tools can greatly increase a child's control. Often, it can be the tool and not the child's ability that is interfering with performance. There are a variety of scissors commercially available and it is extremely helpful to have a selection of them available in the clinic or classroom in order to fit the individualized needs of each child. One disadvantage is that not all styles of scissors are available for both right- and left-handed individuals, with a limited number of styles available for those with a left-hand preference. Conventional scissors range from 4 inches in length for children to 12 inches for adults and are available with both rounded ends for safety purposes and pointed ends for precision cutting. Scissors with clip tips have one wide end for strength and one pointed end for accuracy in cutting. Dick Blick[1] and Nasco[2] carry a wide variety of these in their catalogs.

Fred Sammons, Inc. (Be OK)[3], adaptAbility[4], and Ablewares[5] carry an assortment of adapted scissors for those individuals who lack the strength or coordination to control conventional scissors. Loop scissors, with both rounded and pointed tips, are available from all three companies. These scissors are light weight, have a self-opening handle, and can be used by both right- and left-handed individuals. They are appropriate for individuals with a weak grasp and/or individuals with beginning cutting skills. AdaptAbility also carries a model with finger holes built into the loop, which provides a self-opening mechanism.

Fred Sammons, Inc. and adaptAbility carry speed cutters that have stainless steel blades that spring open and are appropriate for individuals with limited hand and finger function. The scissors come with pointed tips and, even though a plastic case covers the blades, they may be inappropriate for the young child.

Dual-control training scissors are available by Ableware and are appropriate for individuals with poor visuomotor coordination, decreased strength, tremors, poor motor planning, or poor background-foreground discrimination. Blades are rounded and can be used by both left- and right-handed individuals. The handle comes equipped with four loops; the two closest to the blade are for the instructor and the two furthest away are for the learner. The instructor then guides the learner through the necessary steps of cutting.

For those interested in standard school-style scissors, Fiskars[6] offers a sturdy, 5-inch rounded-tip model for less $5.00, which cuts very well, is easy to control, and can be used with either hand. These scissors have a great advantage over regular school scissors in that they cut well and do not dull easily. These scissors can be obtained locally at most hardware stores. Since all children can benefit from using these scissors, it is suggested that they be purchased for an entire class. Therefore, no child would be singled out as having special scissors.

Scissors for the younger child have been designed by Mary Benbow, MS, OTR/L. These scissors are 3 1/2 inches in length with vinyl coated finger loops that facilitate distal placement of the fingers. Successful performance is promoted by its good cutting edge.

For those children who are unable to control any of the above-mentioned scissors due to decreased strength, coordination, or congenital anomalies, electric scissors may be a viable solution. Although inappropriate for the young child, they may provide an alternate method for the older child or adolescent.

Lack of hand preference should not prevent a therapist or teacher from working on scissors skills, but it is recommended that scissors that can be used by either hand be provided. For a child with undetermined hand preference, it can become quite frustrating for the child, parent, therapist, or teacher to have to alternate between left- and right-handed scissors as the child attempts to cut while switching hands.

## Cutting Materials

It is highly recommended that cutting materials be graded according to weight, starting with heavier paper such as oaktag (Klein, 1987). Heavier paper

---

[1] Available from Dick Blick East, Allentown, PA 18105.

[2] Available from Nasco, Fort Atkinson, WI 53538.

[3] Available from Fred Sammons, Inc., Brookfield, IL 60513.

[4] Available from adaptAbility, Colchester, CT 06415-0515.

[5] Available from Ablewares, Pequannock, NJ 07440-1993.

[6] Distributed by Minnetonka, Wausau, WI 54402-8027.

[7] Available from OT Ideas, Inc., Randolf, NJ 07869.

has greater stability, thus allowing the child to concentrate on the control of the scissors rather than the child being concerned with floppy, unstable paper. Klein further suggests graduating paper weight in the following order: index cards, construction paper, paper bags, and eventually regular-weight paper. Later, light-weight material such as waxed paper and aluminum foil can be used when the child can adequately stabilize the paper and cut at the same time. Nonpaper items such as string and fabric are the most difficult to cut and should be reserved until the child has mastered cutting regular paper materials.

## Skill Practice

There are many cutting-skill practice books that are commercially available. They include: (1) *Learning to Cut Series 1-3* (Sieber, 1980), (2) *Learn to Cut* (Wolfe, 1987), and (3) *Pre-Scissor Skills* (Klein, 1987). Each of these programs presents activities in a developmental progression that are graded for difficulty. Programs that begin with activities such as cutting out complex shapes (people, animals, etc.) are designed for children who have already mastered the basic skill of cutting and require advanced activities to further develop skill and promote motivation. In addition to skill practice books, curriculum guides are often available at local instructional support centers or in a school's curriculum library. These curriculum guides provide criterion-referenced assessments along with activities to promote development from one step to the next. Examples of these guides include: (1) *Exceptional Child Education Curriculum K-12: Motor Skills* (Jefferson County Public Schools, 1984); (2) *Project Memphis* (Quick & Campbell, 1985); (3) *Early Independence: A Developmental Curriculum* (Cooke, Apolloni, & Sabbage, 1981).

If you choose to design your own activities, the following suggestions might be helpful. Once the child has mastered the ability to open and close scissors, begin with snipping activities. Long, narrow, strips of paper that require only one snip to cut across are appropriate. Upgrade this activity by increasing the width of the paper to require two snips. As the child progresses, wide cutting lines made with a magic marker will promote eye-hand coordination and control. Grade the activity by slowly decreasing the width of cutting lines. These snips of paper do not need to be discarded. They can be pasted to a piece of paper to form an abstract design or within a designated shape like a large circle or square.

When snipping has been mastered, progress to straight lines, simple shapes (e.g., square), circular lines progressing to cutting a circle and, eventually, figures and designs (Klein, 1987). Depending on the child's vision, bilateral coordination, and endurance, some children will find it easier to graduate from smaller shapes or figures to larger ones while others perform better when progressing from larger to smaller. Make duplicates of your activities to use for pre- and posttesting in order to gauge progress. Klein (1987) and Wolfe (1987) have developed pre- and posttests in their program.

## Adaptations

Holding, manipulating, and/or controlling a piece of paper by the nonpreferred hand may be a problem and can be easily remedied through the use of a tool called "Helping Hands."[8] This device is mounted on a small weighted base that can be C-clamped onto a tabletop. Attached to the base by a ballbearing joint is a bar approximately 3 inches in length, which rotates 360 degrees. At each end, also attached by ballbearing joints, is a pinch clip that can be fastened to each side of a piece of paper. Each ballbearing joint allows for a moderate amount of movement to position the paper while cutting. Wing nuts proximal to each joint secure the joints to maintain needed horizontal or vertical stability (see Figure 3). This tool is also available with a 5-inch crossbar in the Ways and Means catalog and goes by the name of "Twin Grip Fine Work Clamps."[9] It is appropriate for both children and adults. For an individual with no use of his or her nonpreferred hand, it will provide the added stability needed to hold and control the paper. If the individual has some functional use of the nonpreferred hand, it will still permit and encourage involvement

**Figure 3. Child Using "Helping Hands" to Cut Paper**

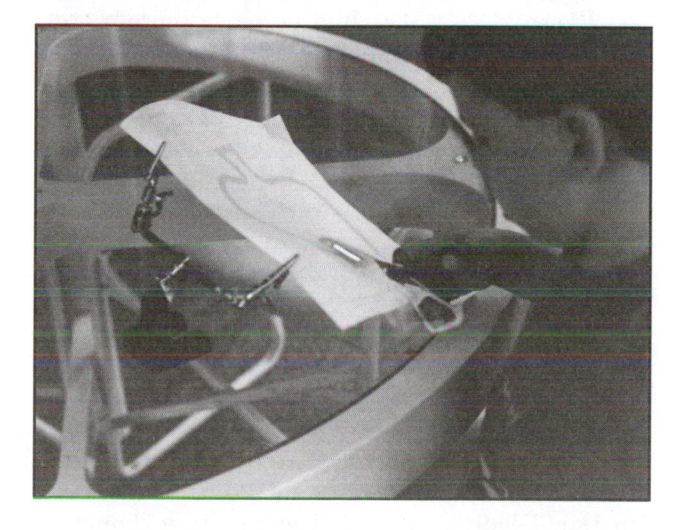

[8] Available from Radio Shack, A Division of Tandy Corporation, Fort Worth, TX 76102.

[9] Available from Ways and Means, Romulus, MI 48174.

**Figure 4. Proprioceptive Feedback Through External Stabilization**

of this hand to hold and rotate the paper. This device works well to stabilize the paper for cutting straight lines with good accuracy.

When incoordination is present, (e.g., tremors or jerkiness) external stabilization by the therapist or teacher at the shoulder, arm, forearm, or wrist may increase control by providing proprioceptive feedback (see Figure 4). As the child becomes more coordinated, the physical support and stabilization can be gradually removed. Tactile input such as yarn or sandpaper guide strips on each side of a cutting line may promote accuracy, however they could also become a visual distraction and therefore should be used with discretion. Often, the use of adaptations is a trial-and-error procedure in an attempt to provide for each child's individualized needs.

### Individualized Education Plans

Educational programs require that an Individualized Education Plan (IEP) be written for each child receiving special services. When writing an IEP regarding a scissors skill, the objective should be written as a functional long-term goal. The objective should be written using behavioral terminology and under what criteria will indicate successful achievement (Klein, 1987). The child's current level of performance should be clearly stated, including any standardized measure used, and the child's sensorimotor status. For measurement of progress, include the methodology to be used, and how the child will be seated. In addition, include specific equipment, the type of scissors to be used, and appropriate activities to reinforce the skill (Klein, 1987).

## Summary

Children with fine-motor deficits often demonstrate difficulty in tool use, including scissors skills. Working with a child to promote the development of scissors skills can be a challenging endeavor for the therapist, teacher, and parent. Very little has been written on the therapeutic interventions appropriate for the child who requires specialized instruction and adaptations to perform successfully in this functional task.

Further research is needed to determine the advantages and disadvantages of specific therapy methods, adaptative techniques, and appropriate scissors selection to enhance treatment planning and curriculum development. Clinical research is also needed for the purpose of standardizing the developmental sequence that is used as a child learns to cut with scissors. This would help in allowing the comparison of the child's scissors skills to his or her overall fine-motor development. Research on the motoric components of arm, hand, and finger control of both the preferred and nonpreferred extremities would assist the therapist in knowing the normal variations so as better to assess the child with fine-motor problems. When this has been accomplished, occupational therapists will be able to effectively treat fine-motor problems, which include scissors skills deficits, with less frustration and trial and error for the therapist, teacher, and the child.

## References


Bleck, E.E. (1975). Cerebral palsy. In E.E. Bleck & D.A. Nagle (Eds.), *Physically handicapped children—A medical atlas for children.* New York: NY: Brune and Stratton.

Brigance, A.H. (1978). *The Brigance diagnostic inventory of basic skills.* North Billerica, MA: Curriculum Associates, Inc.

Bruinicks, R.H. (1978). *Bruininks-Oseretsky test of motor proficiency.* Circle Pines, MN: American Guidance Service.

Campbell, M., & Early, S. (1980). *Cutting with scissors: A study with normal preschool children.* Unpublished manuscript, Boston University.

Coley, I.L. (1978). *Pediatric assessment of self-care activities.* St. Louis: Mosby.

Connolly, K. (1973). Factors influencing the learning of manual skills by young children. In R.A. Hinde & J. Hinde-Stevenson (Eds.), *Constraints on learning.* New York, NY: Academic Press.

Connolly, K., & Dalgleish, M. (1989). The emergency of a tool-using skill in infancy. *Developmental Psychology, 6,* 894-912.

Cooke, T., Apolloni, T., & Sabbage, D. (Eds.). (1981). *Early independence: A developmental curriculum.* Bellevue, WA: Edmark Associates.

Fairbanks, J.S., & Robinson, J.T. (1967). *Fairbanks-Robinson program 1, perceptual-motor development.* Boston: Teaching Resources.

Fishbein, H. (1976). *Evolution, development and children's learning.* Pacific Palisades, CA: Goodyear Publishing.

Folio, R., & Fewell, R. (1983). *Peabody developmental motor scales and activity cards.* Allen, TX: DLM Teaching Resources.

Holle, B. (1981). *Motor development in children: Normal and retarded.* London: Blackwell Scientific Publications.

Jefferson County Public Schools. (1984). *Exceptional child education curriculum K-12: Motor skills.* Louisville, KY: Author.

Karr, M. (1934). Development of motor control in young children: Coordinated movements of the fingers. *Child Development, 5,* 381-387.

Kashihara, E. (1981). Lateral preference and mental ability. *Perceptual and Motor Skills, 52,* 319-322.

Klein, M.D. (1987). *Pre-scissor skills,* revised edition. Tucson, AZ: Therapy Skill Builders.

Larson, C.E. (1976). *Diagnostic and teaching scissors skills.* Niles, IL: Developmental Learning Materials.

Lederer, R.K. (1939). An exploratory investigation of hand status in the first two years of life. *University of Iowa Study on Infant Behavior, 16,* 9-103.

Lopez, M. (1986). *Developmental sequence of the skill of cutting with scissors in normal children 2 to 6 years old.* Unpublished Master's thesis, Boston University.

Lyle, J. (1976). Development of lateral consistency and its relation to reading and reversals. *Perceptual and motor skills, 43,* 695-698.

Marshall, E.D. (1975). Teaching materials for children with learning disabilities. In W. Cruickshank & D. Hallahan (Eds.), *Perceptual and learning disabilities in children.* New York, NY: Syracuse University Press.

McFarland, K., & Anderson, J. (1980). Factor stability of the Edinburgh handedness inventory as a function of test-retest performance, age and sex. *British Journal of Psychology, 17,* 135-142.

Newberg, J., Stock, J.R., Wnek, L., Guidubaldi, J., & Suinicki, J. (1984). *Battelle developmental inventory.* Allen, TX: DLM Teaching Resources.

Petrone, F. (1976). Perceptual motor skills. In *The developmental kindergarten.* Springfield, IL: Charles C. Thomas.

Quick, A., & Campbell, A. (1985). *Project memphis.* Chicago, IL: The National Easter Seal Society.

Rogers, S., & D'Eugenio, D. (1977). Assessment and application. In D. Schafer & M. Moersch (Eds.), *Developmental programming for infants and young children, vol. 1.* Ann Arbor, MI: The University of Michigan Press.

Schneck, C.M. (in progress). Scissors grip patterns in learning disabled children aged 6 to 12 years. Manuscript in preparation.

Schwartz, R.L., & Reilly, M.A. (1981). Learning tool use: Body scheme recalibration and the development of hand skill. *Occupational Therapy Journal of Research, 1,* 13-29.

Sellers, J. (1983). Relationships between hand strength and pencil-scissors skills. *Physical and Occupational Therapy in Pediatrics, 3,* 17-23.

Sieber, L. (1980). *Learn to cut book, 1, 2, 3.* Elizabethtown, PA: The Continental Press.

Smith K., & Sussman, H. (1969). Cybernetic theory and analysis of motor learning and memory. In E.A. Bildeaus (Ed.), *Principles of skill acquisition.* New York, NY: Academic Press.

Stephens, L.C., & Pratt, P.N. (1989). School work tasks and vocational readiness. In P.N. Pratt & A.S. Allen (Eds.), *Occupational therapy for children* (2nd ed.). St. Louis: Mosby.

Stutsman, R. (1948). Mental measurement of preschool children. In *Guide for administering the Merrill Palmer scale of mental tests.* New York: Harcourt, Brace and World.

Welford, A. (1968). *Fundamentals of skill.* London: Methuen.

Wolfe, R. (1987). *Learn to cut.* Tucson, AZ: Communication Skills Builders.

# 7

# Neurodevelopmental Treatment for the Young Child with Cerebral Palsy

*Elizabeth Danella*

*Laura Vogtle*

Traditionally, occupational therapy for children with cerebral palsy has focused on improving their motor and functional abilities. To accomplish these goals, occupational therapists often base their methods on a widely accepted treatment approach, neurodevelopmental treatment (NDT). NDT was introduced by Drs. Karel and Berta Bobath in the 1940s. Incorporating NDT principles into occupational therapy has expanded our knowledge of movement and this additional knowledge has then contributed to our understanding of functional performance. This chapter presents an overview of cerebral palsy, an explanation of the principles of NDT that are important to intervention with a child with cerebral palsy, and a description of treatment planning and intervention with a focus on hand skill development.

## Cerebral Palsy

Cerebral palsy is characterized by sensorimotor dysfunction as manifested by atypical muscle tone, posture, and movement with associated sensory disturbances and/or perceptual deficits (Moore, 1984; Wilson, 1984). It results from a nonprogressive central nervous system (CNS) deficit. The lesion may be in single or multiple locations of the brain and occurs either as a result of intrauterine factors, events at the time of labor and delivery, or a variety of factors in the early development years (Scherzer & Tscharnuter, 1982). Although cerebral palsy is caused by a static encephalopathy, the symptoms often appear to be progressive as the child grows, develops, and attempts to compensate for deficits in motor skill while confronting the force of gravity in an effort to move (Wilson). Moore stated that three of the major functional systems of the CNS may be involved. These are the kinetic (basal ganglion or extra-pyramidal), synergic (cerebellar), and cortical (higher functional systems). Since these systems are interrelated and dependent on one another for normal function, there is generally impairment beyond that of the motor system resulting in sensory, perceptual, learning, and other functional problems.

Children with cerebral palsy have been classified according to type (e.g., spastic, athetoid, ataxic), severity, and distribution (e.g., diplegia, hemiplegia, quadriplegia). However, they frequently demonstrate a combination of patterns and problems (e.g., truncal hypotonia with hypertonic extremities, spasticity with ataxia, or athetoid components). In infancy, it is difficult to diagnose a specific type of cerebral palsy, as the infant often exhibits changing muscle tone and movement patterns during the 1st year. Therefore, close monitoring by an interdisciplinary team of professionals is necessary to ensure optimum development and appropriate intervention for each child and family.

Prior to 1960, the incidence of cerebral palsy remained stable. Epidemiologic studies also indicate little change in the prevalence of children with cerebral palsy born at term (Hagberg, Hagberg, Olow, & von Wendt, 1989). In the 1960s, however, there was a decrease in diplegia seen in preterm, low-birthweight babies due to improved obstetrical management and use of advanced technology in the perinatal centers (Hagberg, Hagberg, & Olow, 1975). Even with these changes, only 50% of infants born at 30 weeks gestation survived (Koops, Morgan, & Battaglia, 1982).

During the 1970s, advances in medical care continued, and medical subspecialties developed in high-risk obstetrics and neonatology. Routine procedures were established for the resuscitation and mechanical ventilation of newborns. Therefore, the survival rate improved. Almost all of the babies born at 34 weeks gestation survived (99%). Even babies born at 26 weeks gestation could live (Koops et al.). With these changes in the survival of smaller, very premature babies, a significant and continuously rising trend in cerebral palsy (spastic/ataxic diplegia) was reported (Hagberg et al., 1989).

Survival continued to improve in the 1980s, as seen in recent data from a multicenter study of neonates born from 1986 to 1987. These centers reported short-term survival rates of 36.5% for babies at 24 weeks gestation and 89.9% survival for babies at 29 weeks gestation (Phelps et al., 1991). Follow-up studies of infants at 2 years of age indicated increased survival rates particularly for the very low birthweight babies (VLBW = 500 to 1,000 grams) (Grogaard, Lindstrom, Parker, Culley, & Stahlman, 1990).

With the increase in survival of these very small babies, medical professionals became concerned that an increase in children with neurodevelopmental handicaps would result. The evidence remains uncertain. Some authors suggest that the increase in survival of healthy VLBW infants has been achieved without an increase in the prevalence of impairment or severe disability (Stewart, Reynolds, & Lipscomb, 1981). However, other studies have documented a significant increase in cerebral palsy in babies born from 1980 to1982 as compared with infants born before the era of ventilator treatment (Hagberg et al., 1989; Kitchen et al., 1986). Grogaard et al. (1990) reported a decrease in the incidence of major and multiple handicaps although the incidence of cerebral palsy remained relatively stable at 7.6% of VLBW infants.

As previously mentioned, because most children with cerebral palsy have functional limitations, occupational therapists have traditionally been members of the intervention team. Although the data are not definitive, clinical experiences confirm the increasing demand for therapists in pediatric hospitals, early intervention programs, school settings, and private practice. The children referred for therapy demonstrate delayed and/or atypical gross- and fine-motor development. Often the child with cerebral palsy has limitations in exploration, play, learning, and self-care that are influenced by the motor delays. Therapists have found that the NDT approach provides a framework for analyzing the child's motor problems and developing an intervention plan to address the child's sensorimotor and functional goals. This approach also defines strategies and techniques for successful implementation of the treatment plan.

## The NDT Approach and Occupational Therapy

Since its inception, NDT has grown and changed, influenced by new understandings of nervous system function and kinesiology (Bly, 1983; Boehme, 1988; Vogtle,1984). This section presents several aspects of the approach that have particular relevance for occupational therapists working with children with cerebral palsy.

### The Relationship of Posture to Upper-Extremity Function

Children with cerebral palsy almost universally have deficits in postural control associated with disturbances in tone and instability of the trunk. As dynamic proximal stability is associated with the child's ability to achieve fine- and gross-motor skill, therapists using the NDT approach emphasize postural control as one of the major goals of intervention. The child has adequate trunk stability when control of the trunk is sufficient to maintain an erect posture, shift weight in all directions, and use rotation within the body axis. By working with the child to gain active upright trunk control and use of active righting and equilibrium reactions, the therapist is establishing a dynamic base of support for more controlled distal function.

Postural control is important to the development of hand skills for several reasons. First, the trunk and arms are integrally connected through the musculoskeletal system. Movements of the trunk and shoulders have a direct effect on distal function. These interactions can be observed in the developmental progress of normal infants. The kinesiological interactions become more apparent when movement develops atypically, and will be discussed in a later section.

Second, a dynamic sitting base is necessary for functional hand use and independence in self-care and school performance. Interaction between trunk control and arm use can be observed in the developmental progression of reaching. Infants initially reach for a target at 4 months of age when their trunks are well-supported. However, when they first attempt independent sitting, their arms are unavailable for reaching. They stabilize their trunk by keeping a wide base of support with both their arms and legs in abduction. They soon learn to bring one and then both arms forward to play while maintaining stability with leg abduction. As they gain control of weight shift and rotation within the trunk, they can easily play in long sitting with the legs more adducted. It is at this time (i.e., 7-8 months of age) that the arms are free for bilateral play activities in independent sitting.

In the past few years, developmental psychologists have begun to recognize relationships between postural development and reaching. Using kinematic analyses, von Hofsten (1979) documented qualitative changes in reaching as infants mature. Using similar analyses, Kamm (1991) suggested that there are interactions between the child's postural skill and quality of reaching. To test this theory, Kamm analyzed the reaches of infants with different levels of postural support. Preliminary evidence indicates that an infant's reaching is smoother with fewer movement units (periods between acceleration and deceleration of movement) when the support level matches his or her postural skill level. The reaches have more movement units (i.e., less-coordinated reaching pattern) when the infant is posturally challenged. These findings support the therapeutic intervention strategy of providing postural support to optimize functional arm use.

Rochat (in press) also investigated reaching in sitters and nonsitters. He found developmental trends from symmetrical two-handed reaching to differentiated unilateral reach as the infant gained postural control for independent sitting. In well-supported positions, the nonsitters would reach with both hands meeting in midline. The sitters, however, used one-handed reaching consistently in all positions. The authors suggest that the infant who does not yet have postural stability has controlled the wide range of possible movements by using the same motion in each hand (bimanual reaching). As the infant gains more postural control, he or she demonstrates more freedom of movement and differentiated unilateral patterns develop.

In another study, Rochat and Senders (1990) researched the coordination between hands, arms, and trunk in reaching while sitting on a mobile platform. The sitters were able to reach and lean forward in synchrony, integrating forward weight shift with reaching. This allowed for increased mobility and larger spaces for exploration. The nonsitters did not shift weight and reaching occurred independently of trunk movement (Rochat & Senders). In a third study, Rochat (1991) analyzed the effects of intervention with 6-month-old nonsitters. When hip support was provided, the infants were able to use a more mature pattern and shift their weight forward while reaching.

## Kinesiological Aspects of Trunk and Arm Function

Analysis of movement using a kinesiological approach has been developed by several NDT instructors (Bly, 1983; Boehme, 1988; Vogtle, 1984). Characteristically, children with cerebral palsy lack the variety of movement seen in normal motor development.

The ability to analyze the sequences of active skills seen in an individual child with cerebral palsy is a useful tool for treatment planning, and can be used in reassessment as a measurement of performance. An occupational therapy perspective will most often focus on the components of action in the upper extremity and those proximal areas that most clearly affect upper-limb movement. These include the spine, head, and pelvis.

The interaction of the upper-limb joints in individuals without disabilities needs to be briefly reviewed before discussing possible limitations in the child with cerebral palsy. A number of authors have contributed to this knowledge over the years (Caillet, 1976; Kapandji, 1970; Norkin & Levange, 1983; Wells, 1967). Describing the biomechanical interactions of the arm in isolation is an oversimplification of the highly complex scope of upper-limb function, but is valuable to understanding the coordination and function of the arm and hand. The brief overview given here is not complete.

The posture of the thoracic spine most immediately influences the shoulder girdle, while the alignment of the thoracic spine is affected by the position of the pelvis and the other areas of the spine. An example of pelvic influence would be the tilt of the pelvis. With a posterior pelvic tilt, the lumbar and thoracic spine tend to be more flexed or rounded, while with an anterior tilt, the lumbar and thoracic spine are extended. Flexion of the thoracic spine facilitates scapular abduction and elevation, while extension allows for a neutral or adducted seating of the scapulae with depression.

The shoulder girdle sets the stage for the functions of the humerus. The range of humeral motion is directly affected by the position of the scapula and clavicle, particularly in overhead and forward-reaching functions. Upward rotation of the scapula and rotation of the clavicle allow the full range of humeral motion overhead in both flexion and abduction. Anterior movement of the clavicle and scapular abduction enable greater forward reach of the humerus into flexion or adduction, while posterior movement of the clavicle and scapular adduction enhance humeral extension for a posterior reach. Scapular adduction also broadens the scope of humeral external rotation; while not actually increasing the range of this movement, scapular adduction moves the glenohumeral joint more posteriorly and complements the range of rotation.

These same sorts of interactions exist more distally in the upper extremity. Humeral external rotation assists in forearm rotation, while humeral abduction and internal rotation are primary pronator assists. Wrist flexion allows greater play in the extensor tendons of the digits for reaching around a larger object, for instance, or covering an octave on a piano keyboard.

The humerus and the elbow control hand placement in relationship to the body (Boehme, 1984; Kapandji, 1970). Humeral flexion at or close to the end of the range and elbow extension are seen with overhead or floor-level activity. Tasks performed behind the body employ humeral and elbow extension. Use of the arms and hand near the torso incorporates small ranges of humeral adduction, flexion, or abduction, accompanied by elbow flexion. The degree of shoulder and forearm rotation shifts frequently during an activity to allow for variation in hand position. The relationship between the wrist and digital activity is highly complex, with the position of the wrist directly affecting the movement sequences of the fingers.

The child with cerebral palsy has abnormalities of tone and muscle control. Contractures and deformities, while usually not present in the young child with cerebral palsy, occur over time as stereotypic movement sequences become habituated and result in soft-tissue shortening with ultimate joint limitations. Therapists working with children with cerebral palsy hope to teach new movement strategies, thus minimizing or preventing deformity and enhancing functional skill. Due to the large degree of variability seen in cerebral palsy, both within distributions of one type, and between types, it is not possible to generalize what compromises cerebral palsy movement. Examples of some specific characteristics follow.

Increased extensor tone is often seen in the child with cerebral palsy, especially in spastic diplegia and quadriplegia. This can result in hyperextension of the head and spine, adduction of the scapulae, with hip adduction and extension, knee extension and plantar flexion of the feet. The upper limbs are often adducted and extended at the humerus, with flexion of the elbows, wrist, and hands. Over time, excessive posturing in humeral extension and scapular adduction can result in loss of soft-tissue mobility, both of the joint capsule, tendons, and ligament structures, and result in loss of scapulo-humeral rhythm. Upward rotation of the scapula occurs after approximately 90 to 100 degrees of humeral abduction or 60 to 70 degrees of forward flexion in the nonimpaired person. When there is tightness between the scapula and humerus, upward rotation of the scapula occurs earlier in humeral motions, resulting in a loss of movement range and functional skill. Tightness between the scapula and humerus often limits the development of active movements of the humerus, specifically external rotation, flexion, and horizontal adduction. Instead, the child uses abduction and internal rotation to achieve movement of the arm away from the body. In the child with cerebral palsy, the inability to isolate motion at one body part contributes to limited movement variability. The child may help move the arm by laterally flexing his or her trunk away from the limb being activated. Internal rotation at the head of the humerus can be attributed to several factors. When the infant first attempts to move the arm, humeral motions are often into extension. The primary extensor of the humerus is also an internal rotator, the latissimus dorsi. With loss of elongation of this and other muscle groups of the rotator cuff (pectoralis major and minor; supraspinatus and teres major), the humerus becomes held in internal rotation. Additionally, the child with a high center of gravity keeps his or her arms close to the body to help maintain an upright posture, further minimizing active range. Elevation of the shoulder girdle, especially when combined with a thoracic kyphosis, also works to maintain the arms close to the sides of the body. For hand-to-hand interaction to take place from this posture, the humerus must be internally rotated.

The functional limitations of the above sequence are limited hand placement in overhead and extended reach, diminished manipulation skills, and difficulty with grooming and hygiene tasks (e.g., hair combing and arrangement, shampooing, hat donning and doffing). Clearly the relationship of muscle control to joint position and alignment affects the child's functional skill.

Characteristic movement sequences within the hand-wrist complex can also be identified in the child with cerebral palsy. Lack of intrinsic control at the digits is common, resulting in limited or absent metacarpalphalangeal flexion, abduction, and adduction and excessive interphalangeal extension. The child compensates by using a combination of assistive efforts at the wrist and the extrinsic muscles of the hand. Wrist flexion is used in the manner described earlier to give greater length to the digital extensors, resulting in hyperextension at the metacarpalphalangeal joints and flexion at the proximal and distal interphalangeal joints. Abduction at the metacarpalphalangeal joints is present, but as a biomechanical result of wrist flexion, not as an actively controlled movement. This compensatory pattern results in marked functional limitations. A power or strong grasp is not possible with the wrist in flexion and with the metacarpalphalangeal joints eliminated from prehension. The size of the object grasped is limited by the lack of full active digital movement. The variety of prehensions available to the individual when cerebral palsy affects the hands is almost always compromised. There is a propensity for deformities of the hand, including web space shortening, hypermobile or subluxing metacarpalphalangeal and carpometacarpal joints of the thumb, swan neck deformities and boutonniere deformities of the digits, contracture, and subluxation of the wrist.

The relationship between functional skill, muscle control, and joint alignment should be clear. From this type of analysis, the effect of muscle control on joint position and alignment can be seen. For treatment planning, an understanding of kinesiological components of function needs to exist in order to identify missing elements that should be incorporated into treatment planning. This aspect of treatment planning will be discussed in detail later in this chapter.

### Facilitating Postural Control to Enhance Hand Function

Although clinical observations support the interrelationships between postural control and upper-extremity function, research is quite limited in this area. In a study on the effects of NDT on reaching in children with spastic quadriplegia, changes in the quality of reaching were documented (Kluzik, Fetters, & Coryell, 1990). After one treatment session that emphasized management of tone during movement and upper-extremity weight shifting and postural reactions, reaches were significantly faster, smoother (fewer movement units), and more mature (Kluzik et al.).

Michaels (1990) also demonstrated a change in the functional performance of two children with spastic diplegia following eight treatment sessions. Of eight outcome measures, the most clinically significant change for both children was the ability to put on a t-shirt independently. Treatment emphasized decreasing tone, increasing trunk and shoulder mobility, and facilitating movement components (i.e., trunk control, weight shifting, overhead reach, and strategies/sequences for putting on a t-shirt).

Although the studies are limited in number, a significant change was demonstrated in performance when postural skills were developed in treatment. Using movement analysis, we can identify problem areas related to tone, mobility, imbalance of movement in the trunk and shoulder girdle, and missing components of movement. Treatment then addresses postural control to facilitate distal skill and modification of tone with the active facilitation of more normal movement components.

For example, a functional goal for a child with spastic quadriplegia might be to catch and throw a large ball with two hands during a group song. Before working on this specific skill, however, other handling techniques might be used to alter tone in the pelvis and trunk so that the child can sit with a symmetrical trunk and a stable base of support during the activity. Then mirror-play activities might be used with the therapist facilitating forward motion and weight bearing into the mirror from the shoulder with the elbows extended. After working on these activities as preparation for the trunk, shoulder, and arm motions, the child might then be more successful with the proximal movement components of the ball play activity. It is important to note that better distal function will not occur simply by working on refining trunk skills. A two-pronged approach is necessary, encompassing both proximal and distal extremity movement in planning therapy activities. Treatment for hand function is discussed later in this chapter and in the other chapters of this book.

A similar analysis guides therapy for a child with spastic diplegia who is working with the teacher on coloring and drawing activities. In assessing his or her movement patterns, the therapist notes trunk and lower-extremity control and posture during gross- and fine-motor tasks. This type of child tends to push his or her upper trunk back and slide his or her hips forward in the chair since the pelvis is tilted posteriorly with the legs often adducted and extended. As a result, the child sits with a narrow, unstable base of support, and the proximal base for fine-motor activity is inadequate. Preparation activities for this child would include handling and/or positioning to bring the hips into more flexion and abduction with an erect pelvis and trunk, thereby improving both sitting posture and balance.

### Importance of Sensation

The Bobaths described cerebral palsy as a sensorimotor problem. They suggested that learning to move is dependent on experiencing the sensations of movement (Bobath & Bobath, 1984). Proprioceptive feedback has a guiding and controlling influence on movement, while the exteroceptors, especially the eyes and the ears, are responsible for the initiation of movement (Bobath, 1980). Since the child with cerebral palsy has abnormal motor patterns, he or she receives abnormal sensory feedback from movement. Therefore, in treatment, emphasis is placed on helping the child experience the sensation of more normal movements through handling and facilitation. Through repetition, these more normal sensations provide a base for the execution of coordinated movement patterns (Bobath & Bobath).

Motor-control mechanisms are now described as using feedback (closed-loop ) and feedforward (open-loop) systems. A feedback, or closed-loop system, uses information generated from the movement itself (Kelso, 1982). Feedback is also used to recognize and correct errors in performance. After a movement is executed, kinesthetic and other feedback is received by a comparison center in the CNS, which compares the feedback from the actual performance with previously stored sensory information associated with that task. Information is then sent back to the muscles so that error correction can be made.

Feedback is important in learning new movement patterns. When an infant is learning to reach, his or her initial movements are poorly coordinated. As he or she experiments and practices movement, it is modified by tactile-kinesthetic feedback from the movement and visual feedback from the environment. During this process, coordination improves and a refined reaching pattern emerges. This process is also instrumental in establishing motor programs, which are sets of prestructured commands that include information on all aspects of the movement (Brooks, 1986).

Feedback mechanisms are emphasized in treatment when varied stimuli and experimentation are offered in an activity. In adjusting responses to varied conditions, motor responses are adapted and movements are refined. When a child is able to climb over a variety of objects, he or she learns to adjust and later modify his or her movement in advance for objects that are large or small, soft or hard. Feedback mechanisms also operate in righting reactions where a stimulus (i.e., weight shift) results in a response (i.e., lateral head and trunk righting). Facilitation of these reactions in therapy may help young children who have not yet developed righting responses to incorporate them into motor programs, thereby giving the child more control of posture and movement. However, it may not help children learn to initiate weight shift in anticipation of movement (feedforward). Although feedback is helpful in learning new tasks, these mechanisms are slow since they rely on learning from previous experience. Therefore feedback alone is an inefficient means for controlling most movements (Bly, 1991; Shapero Sabari, 1991).

In feedforward, instructions for all aspects of the movement plan are prepared in advance and movement is initiated without feedback (Brooks, 1986). Feedback continues to be used to regulate and modify the movement after it has occurred.

Research has shown that feedforward mechanisms control our postural adjustments (Nashner, Shumway-Cook, & Marin,1983). They are responsible for the automatic changes in muscle activity that establish and maintain posture. Feedforward mechanisms organize the temporal-spatial sequence of contractions necessary for weight shift in advance of the actual execution of the task. Cordo and Nashner (1982) documented that postural muscles act before the extremity when subjects were asked to rapidly push or pull a handle. Since postural adjustments use feedforward mechanisms, it is important to incorporate postural responses into play and active, meaningful tasks. The child is encouraged to initiate the postural adjustment and combine it with functional movement. For example, objects can be placed in many different planes and locations so that the child must weight shift in a variety of directions while reaching. Sensory feedback from the weight shift and handling will also reinforce the movement pattern.

Nashner et al. (1983) suggest that poor postural adjustment in some children with cerebral palsy may be primarily related to sensory disorganization in others to motor coordination deficits. They further suggest that sensory disorganization may be related to a breakdown in the feedback mechanism, while motor incoordination problems may be related to the feedforward mechanism.

Clinical examinations of the children with cerebral palsy in Nashner's study also indicated significant sensory and perceptual deficits (Nashner et al., 1983). Therefore, in treatment, we must attempt to determine if a child has sensory- versus motor-based problems. Motor problems are recognizable and are evident when the child does not have the ability to initiate or control the movement pattern. Sensory deficits are more difficult to determine and often require further investigation. The child may show unrefined tactile discrimination skills and poor kinesthetic awareness. Response to vestibular input may also be poorly modulated. These difficulties would give imprecise feedback and thereby affect motor responses.

During treatment, enhanced sensory feedback is used to cue the child as to the body part that should be moved. This can be via deep pressure, joint approximation, or tapping over a muscle belly. Over time and with practice, this use of augmented sensory input is felt to reinforce the feedforward mechanism and enhance anticipatory movement. Activities that increase the child's awareness of the distance and direction of movement and attention to visual monitoring may enhance sensory organization. Motor coordination deficits involve difficulty in planning the postural adjustments, weight shift, and base of support necessary to accomplish a task. Since these components vary with each task, the therapist incorporates weight shift and balance within the functional task rather than working on each component separately. Handling techniques give balanced sensory input to the direction, amplitude, and sequence of movements (Bly, 1991; Shapero Sabari, 1991).

Jeannerod's (1984) study adds further information about the role of sensory information in reach and grasp. In adults, he found that anticipatory hand shaping is seen during the early transport phase of reaching, with some correction made for the size of the object as the object was approached. Since this same pattern was used during three different experimental conditions (with visual feedback, without visual feedback, and without vision), he concluded that hand shaping is part of the motor program for prehension. Although vision refines hand shaping, it does not give information necessary for fine tuning the hand's response to an object, such as weight (Klatzky, McCloskey, Doherty, Pellegrino, & Smith, 1987). This is contributed by tactile/proprioceptive information as feedback once the object is grasped. When visual feedback was removed, the movement pattern was present but accuracy decreased. Tactile-kinesthetic feedback was then important to correct the movement. Therefore, it appears that both feedback and feedforward mechanisms have a role to play for movement to develop in an effective and adaptable manner.

The child's sensory experiences may be limited due to his or her motor dysfunction. Often children with motor problems are unable to tactually explore their body with their hands, (e.g., engage in hand clasping, hand-to-feet, and feet-to-mouth play). When children are not able to move themselves through space, i.e., rolling, crawling, and walking up, down, over, under, through, etc., they do not gain an understanding of their body's relationship to the spatial environment. The child's limited experiences may result in problems in body awareness and motor planning. The importance of these early experiences has also been emphasized by Ayres in her hypotheses regarding the development of motor planning and spatial awareness (Ayres, 1972; Reeves, 1985).

All children with cerebral palsy do not demonstrate these perceptual difficulties. Recent research has demonstrated that children can learn perceptual concepts when motor abilities are limited. Studies in the use of computer technology with children who had limited motor experience indicated that learning occurred through observation even when the child could not imitate or perform the behavior (Behrmann & Lahm, 1984). While learning may occur without the reinforcement of sensory experiences, the importance of experiencing movement and receiving a combination of sensory feedback (i.e., proprioceptive, vestibular, tactile) appears to be critical in developing functional skills such as manipulation.

## Concepts of Inhibition and Facilitation

In NDT, abnormal tone and abnormal movement patterns are seen as integrally connected (Bobath & Bobath, 1984). Although the causal relationship is not supported by research (Bly, 1991; Gordon, 1987), a clinical relationship between tone and movement patterns is seen. One of the basic premises of NDT is that motor function can be improved by changing abnormal tone and movement (Bobath, 1980). The Bobaths state that this is accomplished through "handling the child." During handling, the interaction between the child and therapist is very important. The child guides the therapist and the therapist guides the motor output to make the child's movement reactions as normal as possible (Bobath & Bobath).

These treatment concepts are described as inhibition and facilitation. Inhibition techniques are used to counteract the abnormal patterns while facilitation is used to guide the child's responses. The combination of inhibition and facilitation is used so that both posture and goal-directed movements are as normal as possible (Bobath & Bobath, 1984). Key points of control are used to guide the movement. These key points are areas of the body from which patterns of abnormal activity can be inhibited while at the same time facilitating normal movement (Bobath & Bobath). Proximal key points include the head, trunk, shoulders, and pelvis. Distal key points include the extremities. For example, the arm may be used to facilitate weight shift in the trunk, or the thumb can be a key point of control within the hand. When distal key points are used, the child must control more of his or her own body (Boehme, 1988). When inhibition and facilitation are integrated in handling, the child's active movements are encouraged and movement sequences are facilitated. Inhibition of abnormal movement patterns is incorporated when tone increases during the movement. Treatment is a dynamic interaction between therapist and child. During a treatment session, the therapist may initially handle the child proximally, then progress to more distal points of control and ultimately eliminate manual facilitation. This is certainly the ultimate aim of treatment.

Inhibition and facilitation techniques are based on the effects of sensory input on motor responses (Porter, 1987; Umphred & McCormack, 1985). No sensory system is solely inhibitory or facilitatory. Each has the potential of increasing or decreasing the level of CNS activity depending on the manner in which the sensory stimulus is presented (Porter). Other factors also affect the child's response to sensory input, such as the type of sensorimotor disorder, the child's overall developmental level, and his or her ability to attend to relevant stimuli and inhibit extraneous environmental stimuli.

Inhibition can be attained in a variety of ways. Repetitive and rhythmic movement helps to calm and relax a child as well as decrease his or her tone. As the child's tone decreases, muscles that have been shortened and hold the child in abnormal positions can be elongated. Placing the child on movable equipment, such as balls and rolls, and using activities that require weight shifts can accomplish these goals. Movement over the elongated extremity, such as rolling back and forth over a raised arm in sidelying, will also inhibit scapulo-humeral tightness while elongating the weight-bearing side of the trunk and pelvis. When sitting on a bolster, the child's arm is slowly brought from humeral extension and internal rotation into humeral abduction and/or flexion with external rotation and elbow extension. The bolster may be subtly rocked during the process. Once elongation and inhibition are attained, the arm is brought toward midline to work in a functional space. However, if tone increased during activity, the therapist may return the child to a position in which the muscles with high tone are elongated.

The therapist's touch can be either facilitory or inhibitory and should be graded according to the child's response. Once a comfortable relationship is established between the therapist and child, firm and sustained touch can be used during handling for more inhibitory control. Deep pressure over the muscle belly combined with elongation will also decrease the tone in tight muscles (Boehme, 1988).

In contrast, quick or light touch is facilitory. Therefore, frequent movement of the therapist's hand can be quite disorganizing for a child with athetosis who is easily startled and sensitive to quick movement. Quick touch may be more effective with a low-tone child who needs increased sensory input to respond. For example, light touch may be used along the spine to facilitate trunk extension in sitting when a child is sinking into gravity. More sustained reactions are elicited by repetitive tapping of a muscle group (Boehme, 1988; Porter, 1987). Tapping is frequently used to stimulate the abdominals when a child is first learning to crawl and to facilitate standing with a more neutral or posterior pelvic tilt. Touch during handling is also used to guide the child's movement sequences. This tactile contact can be light or firm or sustained or intermittent, depending on the needs of the child and his or her ability to take over the movement as the therapist releases control.

Symmetrical postures and/or dissociated movements can be used in both an inhibitory and facilitory way. Symmetry is used to inhibit asymmetry and encourage bilateral midline control. However, when a child primarily moves in symmetrical patterns and demonstrates restricted mobility, facilitating dissociated movement and rotation assists in both inhibition of the atypical patterns and facilitation of coordinated movement. When a child has tightness in the shoulders and holds them in an elevated, internally rotated position, tone can be decreased by using firm pressure to bring the shoulders down and back and then moving alternately back and forth (shoulder dissociation). The movement will also increase mobility in the upper trunk. Dissociated shoulder movement can then be actively facilitated as the child bears weight on one arm and reaches forward with the other.

These inhibitory principles are also used in determining appropriate positions and adaptive equipment to maximize function. Midline positioning of head and arms is always encouraged for children with a strong asymmetry. For children with increased lower-extremity tone, a wider base of support with the legs abducted and externally rotated helps to decrease tone and subsequently enhances function. These concepts in positioning are used to promote the child's control during upper-extremity activities, feeding, and other daily activities.

Joint compression elicits a holding contraction, particularly when the joint is in a weight-bearing position (Porter, 1987). It can be given directly through the head and trunk to facilitate an erect posture or through activities such as bouncing on a therapy ball or jumping on a trampoline. Extremity weight bearing is frequently used to assist in maintaining an erect posture and to develop proximal stability in the shoulder and pelvis. Compression through the weight-bearing surfaces can be sustained or intermittent and is graded according to the child's abilities (Boehme, 1988).

Vestibular input is also facilitory especially when it is fast and/or arrhythmic. Inversion also facilitates the vestibular receptors to elicit a postural extension response. This response can be elicited when a child is inverted over a therapy ball and encouraged to push up on extended arms, lift the head, and extend the back. Scooterboard activities and prone inversion must be used carefully in children with cerebral palsy as these tasks may also increase abnormal extensor tone in the lower extremities.

Therefore, positioning of the legs in abduction and external rotation is important to limit this effect as much as possible while the child is actively working with the trunk and upper extremities.

Vestibular feedback is also constantly changing as the child moves and changes position. This feedback is an important consideration in handling as the therapist guides the child in making postural adjustments and movement transitions.

Over the years, the value of combining inhibition techniques with facilitation has become more evident. The child is an active participant in treatment and the therapist can influence tone and facilitate more normal movement patterns during the child's purposeful activity. In this way, carryover into spontaneous movement and function is enhanced.

## Assessment

The focus of this chapter and book is occupational therapy treatment of hand function. While this is the emphasis of this chapter, a brief statement regarding the importance of team assessment and treatment needs to be included. Complex conditions of childhood, such as cerebral palsy, need multifaceted input to encompass the needs of this population. A multidisciplinary team might include some or all of the following: the physicians, (i.e., orthopedic surgeon, neurologist, ophthalmologist, developmental pediatrician); occupational, physical, and speech therapists; orthotist; rehabilitation engineer; educator; parent; dentist; and psychologist. There are other professionals whose services may be needed as well. Communication between the disciplines is key to providing the scope of services necessary for a successful outcome.

The success of intervention depends on the

therapist's ability to assess the strengths and limitations of the child. Since treatment is based on the results of an evaluation, those areas of evaluation most relevant to the NDT approach will be discussed.

Occupational therapy evaluation of the child with cerebral palsy is involved and may cover many aspects of function. The assessment process needs to generate sufficient data for planning an intervention program, selecting appropriate treatment techniques, and judging change.

Many evaluation instruments available to the occupational therapist have limitations or are inappropriate for assessing the child with cerebral palsy. Those tests that are norm-referenced (tests that compare the performance of an individual with a group) do not have special group norms for this population. Most facilities and clinicians develop checklists on their own, which are a collection of clinical observations and developmental sequences.

Available instruments cover functional skills (to a limited degree), developmental level of motor skills, hand function, visuomotor skills, visual-perceptual function, range of motion, and sensory perception status. Which instrumentation the therapist chooses depends on the purpose of the evaluation. In most cases, several instruments will be used. A description of some aspects of assessment follows.

## Functional Abilities

This aspect of treatment is a major premise of occupational therapy philosophy. Function can be described as a variety of tasks, including gait and/or mobility; gross-motor skills such as sitting, reach, prehension, release, manipulative skills; activities of daily living items such as grooming, dressing, feeding, and so on; written and oral communication skills; environmental control; living skills such as meal preparation; and prevocational skills. How a child functions in his or her usual environment (family, community, school) is most important to the occupational therapist.

In spite of the need for functional assessments, there are few available to the occupational therapist. Developmental assessments, e.g., *Peabody Developmental Motor Scales* (Folio & Fewell, 1983) and the *Hawaii Early Learning Profile* (Furono, O'Reilly, Hosaka, Inatsuka, Allman, & Zeisloft, 1985), include items that assess reach, grasp and release, developmental levels of hand skill, and self-care. Elements of manipulation, such as radial-ulnar dissociation, isolation of digital activity, and palmar-digital translation to name a few, are integrated sequences of movements between various parts of the hand necessary for hand function. Until recently, these components of dexterity have been left out of standardized instruments. This area of skill is beginning to be addressed by some authors (Exner, 1990; see Exner, this volume).

Examples of specific functionally oriented instruments that apply to children are the *Pediatric Educational Disabilities Inventory* (PEDI) (Haley, Fass, Coster, Webster, & Gans, 1989) and the *Wee Functional Inventory Measure* (WEEFIM) (Hamilton, Braun, Msall, McCabe, Granger, Kayton, & Goldberg, 1991), based on the adult version, the *Functional Inventory Measure*. Both assess such areas as activities of daily living, general level of communication, transfers, etc. The WEEFIM is designed to assess "burden of care;" it is a questionnaire given to the parents and/or teacher. The PEDI is a criterion-referenced test, also based in part on a parent/teacher questionnaire.

The limitations of current evaluation tools hamper but do not prevent functional assessment. Careful documentation of existing and absent skills can provide the basis for treatment planning. Accompanying videos can helpful to provide greater detail in evaluation and measure progress.

## Quality of Motor Skills

This area is an important aspect of motor performance to NDT. As stated earlier, movement components in proximal and distal skills are felt to be related. Dysfunction in proximal control is hypothesized to lead to inefficient and abnormal distal movement (Bly, 1983; Boehme, 1988).

Quality of movement assessment should be part of an occupational therapy evaluation. Tools that incorporate this as part of a standardized instrument are few. The *Movement Assessment of Infants* (Chandler, Andrews, & Swanson, 1980) is one test that does have quality of movement indicators. One or two fine-motor items on the *Peabody Developmental Motor Scales* (Folio & Fewell, 1983) measure specific movement components, for example, forearm supination. The *Quality of Upper Extremity Skill Test* (QUEST) (Law et al., 1991) has been used in one research study and is currently being used in another. The *Guidelines for OT Assessment* (NDT Course Syllabus, 1991) also include a section on components of movement, as does an NDT clinical assessment (NDT Course Syllabus). These nonstandardized tests, based on clinical observations, indicate that this aspect of performance is an important clinical concern.

## Physical Status

Included in this category are such areas as range of motion, tone assessment, and sensory perception status. As in other aspects of assessment, tools available to the clinician have limited applicability to individuals with cerebral palsy. For instance, normal range of motion for adults is well-documented. However, range of motion in the child differs from that of the adult, which is not generally noted on standard forms. While measurement of range of motion seems straightforward, interrater reliability is low (Riddle, Rothstein, & Lamb, 1987). That is, therapists obtain disparate results when taking goniometric measurements. Reliability is good when the same therapist takes a series of measurements over time (Riddle et al.).

The whole concept of muscle tone, most specifically spasticity, has become an area of intense scrutiny by neuroscientists and health care professionals. The focus on assessing abnormal muscle tone has been prompted in part by the development of the selective dorsal rhizotomy surgery and efforts to document outcomes following this procedure. Currently the *Ashworth Scale*, a somewhat subjective test (Ashworth, 1964; Bohannon & Smith, 1987), is the most consistently used instrument to measure tone in a clinical setting. The search for spasticity measures may ultimately lead to quantification of movement disorders such as athetosis, tremor, and dystonic movements, which are currently not measurable.

While the therapist is hampered by lack of sensitive and accurate measures of tone, clinical judgment of tone is important to comprehensive assessment of the child with CNS dysfunction. Type of tone and its distribution need to be documented. Fluctuations in tone with effort and the impact of these changes on performance need to be described. This is particularly true for hand function, where distal manifestations of tone, such as flexor posturing of the hand, decrease the quality of prehension and manipulation.

Perception of sensation is one aspect of cerebral palsy that remains controversial. A number of researchers have attempted to compare findings on tests of kinesthesia perception between nonimpaired subjects and those with cerebral palsy (Bairstow & Lazlow, 1981; Lazlow & Bairstow, 1980; Nashner, Shumway-Cook, & Marin, 1983; Ophelia-Layman, Short, & Trombly, 1985). Curry and Exner (1988) compared differences in tactile preferences between nonhandicapped children and children with cerebral palsy. The evidence to date seems to support that there are differences between children with and without cerebral palsy, as well as variation between children with different types of cerebral palsy. Other than the measures used in the studies listed above, there are a lack of standardized instruments to measure any of the sensory perceptions. As in assessing muscle tone, lack of accurate instrumentation does not mean sensory perception should be ignored. Clinical observations, made in a systematic manner, can give valuable insights into a child's responses to sensory stimuli. Parental feedback regarding environmental responses can contribute additional information.

### Eye-Hand Coordination/Visuomotor Interaction

These two terms are often and erroneously used to describe the same concept. Visuomotor skill includes but is not confined to eye-hand coordination. It refers to any motor task that incorporates visual input to direct movement output. Other areas of visuomotor interaction include eye-foot coordination and head-activated function, such as headstick use. Eye-hand skill refers specifically to those tasks performed by the hand with visual guidance and monitoring. Both aspects of performance are important to assess in children with cerebral palsy.

Delineating limitations between eye-hand coordination, visuomotor skill, visual dysfunction, visual perception, and purely motoric deficits can be a complex undertaking. The child with cerebral palsy may have difficulties in all of these spheres. It is virtually impossible to separate the interaction of each skill with the others when evaluating performance.

Literally all developmental tests include items that judge eye-hand coordination, and to a lesser degree, visuomotor ability. The eye-hand coordination items are usually included under the fine-motor performance and cognitive subtests. Visuomotor elements are more commonly seen in gross-motor sections of tests. Kicking a stationary ball, walking on a taped line, and jumping over an obstacle are examples of visuomotor activities.

One aspect of assessment not included on standardized instruments is visuomotor coordination in the motor-impaired child when a substitute for the hand is needed. How does a clinician decide the level of visuomotor coordination in a child who is a potential headstick user or who might operate a switch with the chin or cheek? How is performance between these prospective access sites best differentiated? There are computerized assessments that assist the therapist in evaluating which access site is most efficient (Cook, Leins, Harnden, & Zenteno, 1988). Other than these, this component of visuomotor skill must be evaluated using clinical observation and judgment. The areas of assessment discussed to this point should not be viewed as complete by any means; however, those presented are felt to be most relevant to this chapter.

In conclusion, those aspects of performance that the clinician wishes to assess are not fully measurable by the tools available. This state of affairs is slowly changing. In the meantime, careful documentation of skills seen, including video records, is necessary as a basis for planning treatment and judging change.

## Treatment Planning

Treatment planning includes analyzing the results of a comprehensive assessment, plotting the strengths and limitations of the child in question, and setting the short- and long-term goals around which intervention strategies will be organized. Within this process, the concerns of the child, the family, and the needs of the environments in which the child functions are included.

Treatment strategies from an NDT perspective incorporate many issues. Sensorimotor status, including present and absent motor components, integrity of sensory receptors, ability of the child to process and respond to sensory information, presence of mobility limitations, and type and distribution of tone represent some of the concerns to be addressed in treatment. How these items are managed to some degree depends on the developmental and functional skills of the child. There are a variety of techniques available to therapist using an NDT approach, which are chosen because they match the needs of a specific child. For instance, an example cited earlier mentioned a child with increased tone in the hip adductors. A wide base of support in sitting, with the hips in abduction, was suggested as one technique for relaxing or reducing the adductor tone. A wide base of support can be incorporated into a treatment session in other modes than just sitting. Hip abduction can be facilitated during prone activities, with manual guidance and weight shifts, and during standing segments of a session. Manual vibration (rapid oscillation over a selected body area using the hand), dissociation of one hip from another, and facilitation of active abduction and extension of the hips are other treatment alternatives. From this one example, it should be noted that there is not just one solution to management of abnormal tone and movement components. The same solution may not work consistently over time with a given child. As change occurs in the child's responses, alterations in approach are frequently necessary, even within one treatment period.

Facilitation of movement components, building of proximal and distal control, and tone management are tools of NDT treatment that the clinician uses to achieve specified goals. These goals should be stated in functional terms as described earlier in the assessment section. Some settings (schools; certain hospitals) also require measurement criteria, such as duration (time it takes to complete a task), and frequency (number of repetitions successfully performed). This sort of criterion can be useful to a therapist when attempting to assess change. Qualifiers, such as amount of support given or types of cues necessary for goal completion, can be helpful when a child is not expected to be completely independent in accomplishing a certain task.

Many occupational therapy treatment plans are based on task analysis of specific activities. When an NDT approach is used, a task analysis includes both motor and sensory components necessary to complete the activity (see Tables 1 and 2). A volume of information is generated by this type of analysis. The limitations of the child with cerebral palsy are compared to the results of an analysis from a normal child. The identified functional limitations and absent components become the focus of therapy intervention. An NDT approach is used to develop movement and sensory skills, incorporating the techniques of facilitation and inhibition, and tone management. Areas of sensory processing felt to be inadequate can be enhanced by the therapist during treatment in a variety of ways. Hands-on methods, such as manual vibration and tapping, which were described earlier, give differing types of sensory input. The surface used

**Table 1. Posture and Motor Analysis of Reaching for Cup**

| | Step 2 (lifts right arm from lap to table) | Step 3 (right arm reaches for cup; fingers open) |
|---|---|---|
| Head and neck | Both flexed | Same |
| Trunk | Extended; slight rotation to left; right side shortened as arm is lifted onto the table | Slight rotation to left; right side shortened when reaching |
| Pelvis | Neutral and symmetrical | Slight obliquity to the left with reach |
| Hip joint | Flexed to 90°; slight abduction | Unchanged |
| Knee and ankle | Both flexed | Unchanged |
| Shoulder joint | Left adducted at side; right abducted and internally rotated as arm is lifted to the table | Left stays the same; right moves into flexion and midrotation |
| Elbows and forearms | Both flexed to 90° and pronated | Left unchanged; right moving to extension and midposition at forearm |
| Wrists, hands | Wrists slightly extended; all MPs and IPs flexed; thumb extended and adducted | Left unchanged; right wrist extended and radially deviated; MPs extending; IPs slightly flexed; thumb abducted |

**Table 2. Sensory Analysis of Reaching for a Cup**

| | Step 1 (right arm is lifted from lap to table) | Step 2 (right arm reaches for cup; fingers open) |
|---|---|---|
| Visual feedback | May be used to locate cup | Isolates cup, then looks away |
| Audition | Not used | Not used |
| Proprioception | Feedback as to position in space while sitting | Same |
| Kinesthesia | Input as to head and neck position; also arm position change | Input as to arm forward motion and finger extension |
| Vestibular | Used in balance and upright posture | Same |
| Touch | Arm may brush table top | Touch of cup as fingers reach to it |
| Deep pressure | Area of buttocks and back touching chair | Area of buttocks and back touching chair |
| Two-point input | As above | As above |
| Temperature | Not a factor | Not a factor |
| Pain | Not a factor | Not a factor |

during treatment (bolster, ball, suspended apparatus) also provides the child with diverse sensations, as does the activity chosen for the child to perform.

The step-by-step task analysis serves as a basis for goal setting, and as a measurement tool to assess progress in treatment. The quantification measures discussed earlier (duration, frequency, and achievement) are assigned to a step or certain series of steps of the task. Successful completion of a specified step indicates it is time for new goal.

## The Intervention Process: Developing Hand Skills in Children with Cerebral Palsy

When developing hand function in the child with cerebral palsy, the clinician considers the individual child's abilities, his or her movement patterns, and limitations to function. Once these have been identified during the assessment, intervention strategies can be developed to address the components specific to the child's performance. Treatment is complex, integrating inhibition and facilitation techniques with task analysis of age-appropriate activity to develop missing components and build function. The NDT approach evaluates proximal control and its influence on distal skills prior to isolating distal function concerns. For example, trunk alignment and stability are emphasized as one aspect of intervention. Sustained upright posture combined with graded weight shifts and rotational freedom provide the dynamic control and stability needed for the child to achieve both large-body movements and complex dexterity skills. As mentioned earlier, this does not mean that treatment of wrist and digital dysfunction should be ignored until proximal goals are reached during therapy sessions. An approach where proximal and distal concerns are simultaneously incorporated into treatment needs to be used.

## Intervention Strategies

### Techniques for Tone Modification

Increased tone is modified to decrease its impact on active movement. The preferred mode of achieving distal tone change is by modifying tone in the trunk and shoulder girdle. If this does not adequately diminish tone for function, the problem may need to be addressed directly within the hand. Manual vibration, applied to a central area or directly over the body part where hypertonia is interfering, and traction to the joints involved are used to inhibit increased tone, as are graded degrees of upper-extremity weight bearing. Traction refers to the application of a sustained gentle pull directly below a joint that is supported from directly above the articulation. Hand shaping, a technique in which the various arches of the hand are manually approximated using the therapist's fingers to accentuate their shape, can also be used for tone reduction as well as an aid to mobility within the carpals and metacarpals of the hand. These techniques can result in a hand with diminished tone. Approximation is a technique in which the two sides of the joint are supported and aligned, then compressed gently toward each other. Often traction/vibration are used alternatively with approximation as needed to achieve and maintain a more normal tone status.

The thumb is a key point of control in the spastic hand. Elongation of the web space with abduction and extension of the thumb at the carpo-metacarpal joint is accomplished by using firm pressure on the volar surface from the base of the first metacarpal along the thenar eminence. Traction applied to the first metacarpal can further reduce tone. Active motion is then facilitated. Movement sequences within the wrist-hand complex are intricate. Increases in tone and associated reactions are manifested quickly with effort. Tone management techniques need to be reapplied periodically when tone fluctuations are noted.

## Upper-Extremity Weight Bearing

Upper-extremity weight bearing has been used in NDT as a method of inhibiting tone within the hand while elongating the wrist and finger flexors. It can also be a valuable tool for applying deep sensory input within the hand, wrist, elbow, and shoulder. Lateral weight shifting while maintaining weight-bearing positions results in proprioceptive input and deep touch across the palmar surface, which facilitates radial-ulnar dissociation in the hand. Anterior-posterior weight shifts allow for this same mobile sensory input moving over the volar surface of the hand from the thenar eminence to the tips of the fingers. Care needs to be taken during weight bearing that the child is using active control of the upper limb and trunk rather than stabilizing on hyperextended elbows. Frequent weight shifting during weight bearing rather than sustaining a static position activates both trunk and arm musculature. The therapist employs endless variations of weight-bearing activities. The child can weight bear sitting on equipment with arms positioned posteriorly in extension and external rotation or anteriorly in forward flexion and midway between internal and external rotation. Another variation of weight bearing would have the child lying prone over the ball, resting his or her hands in the therapist's hands. In this manner, the amount of weight through the arms and hands can be controlled, as can the degree and range of weight shift. Gradual increases in the child's range of wrist and finger extension can be facilitated by the therapist when limitations exist.

The clinician needs to remember that weight bearing is fatiguing. Movement transitions give the opportunity for moving over the upper limbs when traversing from one position to another. Additionally, weight bearing is only one aspect of upper-extremity treatment. It should be used in combination with other interventions to promote specific functions in the upper extremities and hands.

## Stability and Mobility: Concept and Practical Application

The hand has many moving parts that are used in rapidly changing positional variety during dexterous activity. The intricacy of manipulative skill can be overwhelming to the clinician attempting to plan treatment strategies to improve such function. A step-by-step plan that interchanges the stability and mobility functions of the hand during interdigital activity is important. Stability refers to the maintenance of a posture or motion. Mobility indicates active motion within a hand or upper extremity. Most manipulation within or between the hands uses aspects of both components. For instance, in-hand manipulation includes mobility on the radial side of the hand to open the hand, radial stability during grasp, radial-ulnar mobility during transfer, and ulnar stability during storage on the ulnar side. To pick up another object, the radial side once more activates in-hand opening and stabilizes for grasp, with radial and partial ulnar mobility for transfer. The ulnar side, during radial activity, is holding or stabilizing the first object grasped. Since an object remains in the hand while another is being transferred into the ulnar side, one or two ulnar fingers grasp the original object while the new object is moved one finger at a time.

This concept, while perhaps confusing, is important to use while building new hand skills. The therapist, using hands-on control, provides stability in one area, while requiring either stability or mobility in another area. For example, when facilitating active metacarpalphalangeal extension, the therapist may stabilize the wrist in a neutral or slightly extended position. This allows the child to concentrate on the dynamic or mobility skill, while the stability is being provided by the therapist. Over time, the therapist gradually withdraws control at the wrist, asking the child to provide the control for both stability and mobility. In another example, the therapist may seek to facilitate isolated digital function. The stability is provided for the child by the therapist, sustaining all but the active digit in flexion, while the child activates the mobile finger into extension.

In certain types of cerebral palsy, most notably athetosis, the child may be taught to provide his or her own proximal stability to allow for more distal control. When working on computer use, the youngster with athetoid involvement may be encouraged to lean on his or her elbows for stability. This may facilitate better distal control of the hand in these tasks and give the child greater independence in activities with family and peers. Facilitating or assisting proximal control of the trunk and movement of the shoulder remains part of treatment, but is used during preparation for function or during a skill that requires less distal mobility. In this manner, treatment is an ongoing melding of stability and mobility between and within segments of the upper extremity.

## The Use of Activity

Although the above ideas for handling have been presented as specific techniques, they are integrated into motivating age-appropriate activities related to the child's functional goals. While the Bobaths actively involved the child in the treatment process, the emphasis was on facilitating more normal movement. Voluntary effort, especially with power, was discouraged because it was felt to increase spasticity and produce associated reactions (Bobath & Bobath, 1984). As more is understood about the motor learning process, this concept is being modified. Now more emphasis is placed on the value of goal-directed activity (Bly, 1991). When new movement sequences are being facilitated, the child's active participation is needed to use the new motions in problem solving and performing goal-oriented tasks so he or she can develop new motor programs. The quality of movement must be balanced with the benefits of having the child participate in planning, initiating, and producing movement. The skill and art of treatment are in recognizing when and what motion components to compromise in order for the child to solve motor problems as he or she learns new motion sequences. As noted earlier, the therapist, when planning treatment, analyzes the child's functional abilities, determines the movement and sensory components needed for a specified functional goal, selects activities to move toward the goal, structures the environment to elicit goal-oriented problem solving, and guides the child's movement during these activities. The child's age, motivation, and response to treatment, in combination with the specified goal, will determine what play activities will be used, and what modifications in the environment need to be made. The functional goal itself may be the activity used, for example, self-feeding or computer use. To better engage children of a certain age level, the therapist or a doll could be "fed" by the child. Dress-up with adult clothing can be fun, and often provides larger-size buttons for fastener practice. Play can also be graded to build specific components. There are many nonprehensile activities, i.e., finger painting, clapping games, and smearing shaving cream, that would provide one medium for promoting finger extension to promote active release. This type of engagement provides generalized sensory input that can be graded by texturing the materials, i.e., adding flour or other heavy substances.

Environmental modifications include such items as the height of the surface where the activities are placed, the position of the activity, the size of items used in a given play selection, textural modifications to a surface, and so on. This type of adjustment can be key in facilitating certain movement elements. For instance, hand function carried out on a vertical or angled surface is more likely to promote wrist extension, while a horizontal surface more often encourages wrist flexion. The same vertical surface in combination with the body positioned arm's distance away, inspires elbow extension and humeral flexion. Rather than asking a child to reach and take an object, the therapist can ask the child to hold a hand out to receive the object. This type of request facilitates forearm supination. The therapist could then hold out his or her hand for the object to be returned, encouraging end range movements between pronation and supination in the child.

## Specific Elements of Hand Function

In this section, examples of specific hand functions, reach, grasp, release, and interdigital interactions will be briefly presented, with examples of how to facilitate each.

### Reach

Reach exemplifies the influence of proximal stability on distal control in its broadest sense. Dynamic trunk control is necessary for reach to occur in a variety of fields (overhead, down to the floor, off to the side). While the focus of this book is specifically on hand function, it is helpful to point out that reach can be used to activate broad and graded ranges of trunk motion as well as ranges of shoulder motion. Some children with spastic diplegia have better glenohumeral control than postural control, for example, and would need to refine trunk skills rather than the shoulder aspects of reaching.

For functional reach, the child should also have adequate range in the shoulder girdle, glenohumeral joint, and elbow. If range is limited, it is the first priority for the therapist to address. Areas where limitations often occur are the upper chest, lack of dissociation between the scapula and humerus, and diminished scapular abduction/adduction. Inhibition of tone in the trunk and shoulder girdle combined with facilitation of desired motion may eliminate tightness, although in some instances, specific soft-tissue mobility needs to be incorporated. For example, muscle shortening is commonly seen in the pectoralis major and minor. Deep pressure applied over the belly of the muscle combined with elongation can be used to improve mobility. To actively reinforce the increased range, reach to the side and then behind the body is required in an activity. Weight shifts in the pelvis and lateral mobility and upright control in the head and trunk are necessary, as well as active humeral abduction and a range of humeral rotation, depending on where the reach is aimed. As reach is required more posteriorly, clavicular and scapular movements change. The scapula must be more adducted for reaching backward, while clavicular motion at the sternoclavicular joint needs to be in a horizontal plane and posterior direction. A wide range of trunk rotation further facilitates reach behind the child. Alternating reach to the front with reach behind can increase the range of trunk rotation in the child with limited mobility. Controlled rotation in midranges can be facilitated in children with athetosis or ataxia.

When building components of reach, it is not necessary, and frequently not desirable, at first, to incorporate grasp. This depends on the skills of the child. If the effort of movement produces associated reactions in the hand, then the activity used to facilitate the reach should encourage finder extension, e.g., pointing, hand slapping, or pressing the hand into paint in front of the child and turning to make hand prints on paper taped to the wall in back of the child. The goal over time obviously is to integrate grasp, then possibly bilateral grasp, with reach. There are also grooming tasks, such as hair arrangement, that use reach, grasp, manipulation, then release with reach and sustained arm posture away from the body. All of these components of function would be listed when the task analysis discussed earlier is performed.

## Grasp

Refined prehension is dependent on the coordination of forearm, wrist, and hand motion. As discussed earlier, atypical combinations of movement in the shoulder complex often lead to characteristic hand-wrist postures and limitations in hand function. Deficits of hand function can also exist in the child who does not demonstrate the marked shoulder dysfunction presented at the beginning of this chapter. The child with mild hemiplegic involvement may be an example. In either case, proximal treatment of the trunk and shoulder girdle will not eliminate the problems that exist within the hand. Specific treatment activities at the forearm and hand are also necessary. The therapist uses handling techniques to alter tone, increase or moderate mobility, and facilitate isolated movement of the forearm, wrist, and hand. These motions are then combined to produce various types of prehension. The therapist facilitates radial-ulnar dissociation and isolated finger control as the foundation for intricate dexterity skills and for prehension in all of its variety.

Tone management within the hand as presented earlier in this chapter applies to both prehensile and manipulative elements of hand function. Careful monitoring of tone during very specific treatment activities, such as prehension, is needed. Intervention is often reintroduced to manage the tone increases that occur when the child applies increased effort.

Range of motion in the hand includes mobility of the carpals and metacarpals, which provide the structure of the hand arches. These structures contribute to the breadth of grasp and are a factor in radial-ulnar dissociation. The three arches—the palmar transverse arch, longitudinal arch, and distal transverse arch—are key structural and functional aspects of the hand that need to be considered and addressed during treatment. Hand shaping has been presented as a method of reducing tone within the hand. It is also a technique for improving the passive mobility within the arches of the hand, and for giving sustained deep pressure input across the arches. The other important factor in developing and maintaining the palmar arches is active muscular control. Flattening of the arches occurs when extrinsic control of the digits and wrist flexion is used exclusively. Therefore, treatment of the hand that focuses on developing more normal balance between intrinsic and extrinsic musculature contributes greatly to developing and sustaining the structure of the arches.

Mobility and range issues are the most basic issues to be considered in hand treatment. The type of prehension to be developed over a session or series of sessions needs to be considered also, as do the differing kinds of digital activities incorporated in the prehension chosen. Objects to be grasped are carefully chosen. Size, texture, weight, and type of play activity the object can be used for are key factors, as are potential modifications of the object as the child's abilities change with treatment.

Suppose the child in treatment is a 6-year-old with athetoid involvement, quadriplegic distribution. The short-term goal for the session is that she will sit supported in a chair and grasp a can (modified as a salt shaker) using a cylindrical prehension (incorporating the thumb, a step up from her usual palmar prehension) and shake out salt within a certain time span, needing only one try to successfully pick up the can. The activity for the day is making salt clay, and the child is to add the salt when indicated by the therapist. The can is set into a 1-inch-deep tray with a round cut-out to stabilize the can against extraneous motion. In preparation for the prehensile task, the child sits on a bolster and walks her hands over hand prints taped to the wall in a variety of directions. To prepare the forearm for changing from pronation to midposition, some of the hand prints have been positioned in a sideways manner. Joint approximation has been applied to each MP joint of the hand being used for prehension, and to the wrist of the opposite hand that she will attempt to stabilize on the table. Extra time for joint approximation has been given to the CMC joint of the thumb on the grasping hand, since the prehension required involves thumb abduction. To prepare for the activity, the child has positioned the bowl in which the clay is to be mixed on the table, with the therapist providing considerable stability and control at the elbow and forearms. The child adds salt a few shakes at a time, giving the child a number of opportunities to pick up the can. The therapist provides control at the wrist during the first few prehensions, with some deep pressure to the palm of the hand to accentuate the shape of the hand desired and to facilitate thumb abduction. As the child's thumb activation begins, the deep pressure cue is eliminated. Control at the wrist can later be moved proximally up to the elbow if possible. Continued reapplication of deep pressure can be given to the palm of the hand if thumb placement is not consistent.

The example cited points out the importance of using sensory input judiciously. The approximation was to build tone and awareness of position in space, as was the weight-bearing activity. The deep pressure into the palm of the hand was to facilitate thumb motion and to activate the flexors of the fingers. By applying pressure to the palm, with the wrist controlled in extension, the flexion facilitated would be IP as well as MP flexion, incorporating both interossei and lumbrical muscle groups.

The concept of providing stability to allow for distal mobility was demonstrated here as well. Stability was provided at the wrist and forearm at different times. If distal control was adequate, stability was withdrawn. When needed, it was reapplied.

Developing prehension requires careful analysis of the movement components needed, the problems currently being encountered by the child, and the issues of dysfunction to be addressed in therapy. Based on the identified needs, specific methods and engaging activities are designed.

## Release

The hand's release of an object encompasses digital extension in a controlled manner, hand placement for release, planning and organization of the motor components, as well as cognitive and perceptual awareness of the appropriate timing of release. The active elements of release will be discussed here.

Refined release incorporates the interossei musculature as well as the long finger extensors. Extensors and possibly abductors of the thumb are needed as well. The child with increased tone may lack the intrinsic control necessary, so uses wrist flexion in combination with the extensor digitorum to provide release. This movement pattern results in an ungraded gross release that is inefficient. Some children with hypertonia may be unable to release items at all. The child with athetosis, whose hand may be very mobile, tends to use ungraded finger extension that results in rapid and uncontrolled release. This can be complicated by extraneous motion in the upper limb. Depending on the situation, the therapist designs activities to assist the child in developing finger extension, both of all the digits simultaneously and individually, and in combination with a variety of wrist and forearm positions.

Preparation for working on the volitional skill of release, as with prehension and reach, needs to be carried out carefully, as in the following example. A 2-year-old boy with hypertonus, quadriplegic involvement, and normal intelligence is able to reach and grasp with great effort, but releases only by "scraping" objects against other surfaces until they fall out of his hands or by raising his arm as high as he can and shaking his trunk and arm. The goal in therapy is that he will sit with support straddling a bolster at a table and demonstrate unilateral digital extension of all fingers and thumb when placing his hand on a plate switch to activate a tape recorder. As preparation, a large ball is used. The child is placed in prone over the ball, with lower extremities dissociated (i.e., with one hip flexed and one hip extended). His upper extremities are supported in overhead flexion by the therapist, whose hands are midway between the child's shoulder and elbow. A clapping game to music is used, with the therapist providing the force for clapping. The ball is moved simultaneously in a side-to-side direction. Manual vibration is given by the therapist between the shoulder and elbow if tone increases are felt at the shoulders. The game is used to decrease tone and provide sensory input to the palms of the hands, as well as to engage the child. From here the child "climbs" up to a sitting position with considerable assist from the therapist. The shoulder continues to be the key point of control. This activity provides upper-limb weight bearing to sustain digital extension on the ball's surface with the effort required to push into sitting. Several excursions of this nature are performed to each side, then the therapist positions herself in back of the child, who is sitting on the ball. The plate switch is introduced and demonstrated. The surface of the switch may be altered or textured to provide more sensory input if desired. The child is encouraged to use it, with instructions to attempt to keep his hand flat. Support is provided at the palm, with deep pressure given in a down and back direction on the dorsum of the hand. If efforts to activate the switch result in digital flexion, manual vibration at the palm is provided.

As indicated in the example, the preparatory nature of the therapy session is instrumental in the successful outcome desired. As with other components of hand skill, once volitional extension is achieved, the next step would be to incorporate the motion into the skill of release.

## Manipulation

This highly complex phenomenon is not just one action, but a whole series of interactions among the various digits and the thumb. For clinicians attempting to remediate manipulative dysfunction, the intricacy of the movements can be overwhelming. Careful task analysis, as discussed earlier, can be particularly useful. The finger movements that are involved in a task should be identified to facilitate treatment planning. The importance of the intrinsics in hand activity has already been emphasized, particularly regarding flexion and extension of the joints of the fingers. Other movements that are significant to moving an object within the hand are the lateral motions of abduction and adduction and ulnar deviation movements.

In addition to the specific movements needed, the sequencing of activity is also important. For instance, radial rotary interactions involving the thumb, index, and middle fingers incorporate alternating actions in a linear order. The ability to organize this progression is necessary to perform the task, while reversing the order from clockwise to counterclockwise needs to be present for efficiency. Another example of this is radial-ulnar directional motions. To store objects during in-hand manipulation, items are moved from the radial side of the hand to the ulnar side. To retrieve the items, ulnar-to-radial action is necessary. The capability to sequence reciprocating actions, and in a variety of ways, is another important component of manipulation.

The sensory aspects of manipulation are as critical as the motor elements. During in-hand and bimanual manipulation, the fingers respond rapidly to subtle sensory input. When planning treatment, the clinician needs to know how a given child responds to different types of sensation. If a child has an aversion to a sensory mode used in treatment, it will not facilitate the desired outcome. Sensory input is graded during treatment to provide deep pressure input at first. This is gradually decreased to lighter sensation. The child with cerebral palsy needs to be able to respond to less dense stimulus if he or she is to demonstrate manipulative proficiency.

An example should help to clarify an NDT approach to manipulation. The child is a 7-year-old girl with a mild right hemiparesis. She has spontaneous assistive use of the right arm. Reach, grasp, and release are present. Types of grasp available are palmar and cylindrical, lateral pinch, and a modified pincer grasp to the middle finger. The treatment goal is as follows: The child will be able to use the first three digits of the right hand to access a computer keyboard for word processing, using the thumb to hit the space bar, and the index and middle fingers to access keys on the right hand side of the keyboard. Currently she sustains the wrist in flexion and ulnar deviation, causing frequent errors in finger placement of the right hand. In this situation, some of the treatment session would need to be devoted to developing active wrist control. Preparatory activities include standing and pushing a large therapy ball against resistance provided by the therapist, and in rolling the ball in a hand-to-hand sequence moving to both right and left. These preparatory sequences both use elbows and wrists extended simultaneously, and incorporate dynamic sequencing, first bilaterally in an anterior-posterior direction, then in a lateral direction with reciprocating motions. The next activity focuses on isolated finger motions. A series of finger plays, which use specified finger motions to accompany rhymes, are demonstrated by the therapist. After demonstration, the therapist takes the child's fingers through the sequence, providing joint approximation in the process. If assistance is needed to stabilize flexed fingers, it is provided. Wide-range reciprocating finger motions, with isolated finger function, are built using this activity. Extra support is given at the wrist, if needed. Keeping the elbows extended helps to facilitate wrist extension, while using the radial side fingers facilitates radial wrist extension. If associated reactions are seen in the right arm, they are addressed using manual vibration or tapping over the muscle belly. This child frequently pulls into elbow flexion, hence tapping over the belly of the triceps is used to facilitate elbow extension. The therapy session is completed with actual computer work. The computer program chosen is a game that uses the space bar and arrow keys. The child's right index finger is used on the arrow keys, with the thumb used on the space bar. The keyboard is set back from the edge of the table to allow the forearms to rest on the surface. Control is provided at the wrist when needed.

Dexterity skills are the highest-level abilities available in the hand. There are children with cerebral palsy who will not attain this degree of function. The more mildly affected child does have the potential to develop some competence in this area.

## Conclusion

The aim of this chapter has been to describe the NDT approach and its application to specific hand function treatment. The therapist uses the NDT framework to promote development of hand skills by considering the child as a whole person. The influence of each body part on others is taken into consideration and factored into treatment planning. The neurophysiological basis for NDT is changing as new information is uncovered. For example, the treatment framework has shifted to emphasize functional outcomes. The NDT treatment perspective continues as a powerful tool in the clinician's repertoire. The role of NDT in hand treatment is continuing to develop as knowledge in this intricate skill expands.

## References

Ashworth, B. (1964). Preliminary trial of carisoprolol in multiple sclerosis. *Practitioner, 192,* 540.

Ayres, A.J. (1972). *Sensory integration and learning disorders.* Los Angeles, CA: Western Psychological Services.

Bairstow, P.J., & Laszlo, J.I. (1981). Kinesthetic sensitivity to passive movements and its relationship to motor development and motor control. *Developmental Medicine and Child Neurology, 23,* 606-616.

Behrman, M.M., & Lahm, L. (1984). Babies and robots: Technology to assist learning of young multiply disabled children. *Rehabilitation Literature, 45,* 194-201.

Bly, L. (1983). *The components of normal movement during the first year of life and abnormal motor development.* Oak Park, IL: NDT, Inc.

Bly, L. (1991). A historical and current view of the basis of NDT. *Pediatric Physical Therapy, 3,* 131-135.

Bobath, K.A. (1980). *A neurophysiological basis for the treatment of cerebral palsy* (2nd ed.). Philadelphia: Lippincott.

Bobath, K.A., & Bobath, B. (1984). Neuro-developmental treatment. In D. Scrutton (Ed.). *Management of the motor disorders in children with cerebral palsy* (pp. 6-18). Philadelphia: Lippincott.

Boehme, R. (1984). Shoulder girdle stability as a basis for fine motor activity in the first six months of life. In *Selected proceedings from Barbro Salek memorial symposium* (pp. 163-168). Oak Park, IL: NDT, Inc.

Boehme, R. (1988). *Improving upper body control.* Tucson, AZ: Therapy Skill Builders.

Bohannon, R.W., & Smith, M.B. (1987). Interrater reliability of a modified Ashworth scale of muscle spasticity. *Physical Therapy, 67*(2), 206-207.

Brooks, V.B. (1986). *The neural basis of motor control.* New York: Oxford University Press.

Caillet, R. (1976). *Hand pain and impairment.* Philadelphia: F.A. Davis.

Chandler, L., Andrews, M., & Swanson, M. (1980). *Movement assessment of infants.* A. Larson (Ed.). Rolling Bay, WA: Authors.

Cook, A., Leins, J., Grey, T., Harnden, S., & Zenteno, C. (1988). *The switch assessment program for the Apple IIE* [computer program]. Sacramento, CA: The Assistive Device Center, California State University.

Cordo, P.J., & Nashner, L.M. (1982). Properties of postural adjustments associated with rapid arm movements. *Journal of Neurophysiology, 47,* 287-302.

Curry, J., & Exner, C.E. (1988). Comparison of tactile preferences in children with and without cerebral palsy. *American Journal of Occupational Therapy, 42*(6), 371-377.

Exner, C.E. (1990). The zone of proximal development in in-hand manipulation of skills on non-dysfunctional three and four year old children. *American Journal of Occupational Therapy, 44*(10), 884-891.

Folio, M.R., & Fewell, R.R. (1983). *Peabody developmental motor scales and activity cards.* Allen, TX: Developmental Teaching Resources.

Furuno, S., O'Reilly, K., Hosaka, C., Inatsuka, T., Allman, T., & Zeisloft, B. (1985). *Hawaii early learning profile.* Palo Alto, CA: VORT Corporation.

Gordon, J. (1987). Assumptions underlying physical therapy intervention: Theoretical and historical perspectives. In J.H. Carr, R.B. Shepherd, J. Gordon, A.M. Gentile, & J.M. Held. *Movement science: Foundations for physical therapy* (pp. 1-30). Gaithersburg, MD: Aspen Publishers.

Grogaard, J.B., Lindstrom, D.P., Parker, R.A., Culley, B., & Stahlman, M.T. (1990). Increased survival rate in very low birth weight infants (1500 grams or less): No association with increased incidence of handicaps. *Journal of Pediatrics, 117,* 139-146.

Hagberg, B., Hagberg, G., Olow, I., & Wendt, L. von. (1989). The changing panorama of cerebral palsy in Sweden, V. The birth year period 1979-1982. *Acta Pediatrica Scandinavia, 78,* 283-290.

Haley, S., Fass, R., Coster, W., Webster, H., & Gans, B. (1989). *Pediatric educational disabilities inventory (standardization version) (PEDI).* Boston, MA: New England Medical Center.

Hamilton, B., Braun, S., Msall, M., McCabe, M., Granger, C.V., Kayton, R., & Goldberg, R.M. (1991). *Functional independence measure for children (WeeFIM).* Buffalo, NY: SUNY Research Foundation.

Hinderer, K., Richardson, P., Atwater, S. (1989). Clinical implications of the Peabody Developmental Motor Scales: A constructive review. *Physical and Occupational Therapy in Pediatrics, 9*(2), 81-106.

Hofsten, C. von. (1979). Development of visually directed reaching: The approach phase. *Journal of Human Movement Studies, 5,* 160-178.

Jeannerod, M. (1984). The timing of natural prehension movements. *Journal of Motor Behavior, 16,* 235-254.

Kamm, K. (1991). *Postural influences on reaching.* Conference presented by Educational Seminar Associates, Newark, NJ.

Kapandji, I.A. (1970). *The physiology of the joints, Vol. 1 upper limb.* New York: Churchill Livingstone.

Kelso, J.A.S. (1982). Concepts and issues in human motor behavior: Coming to grips with the jargon. In J.A.S. Kelso (Ed.), *Human motor behavior: An Introduction* (pp. 21-58). Hillsdale, NJ: Erlbaum.

Klatzky, R.L., McCloskey, B., Doherty, S., Pellegrino,

J., & Smith, T. (1987). Knowledge about hand shaping and knowledge about objects. *Journal of Motor Behavior, 19,* 187-213.

Kluzik, J., Fetters, L., & Coryell, J. (1990). Quantification of control: A preliminary study of effects of neurodevelopmental treatment on reaching in children with spastic cerebral palsy. *Physical Therapy, 70,* 65-78.

Kitchen, W.J., Rickards, A.L., Ryan, M.M., Ford, G.W., Lissenden, J.V., & Boyle, L.W. (1986). Improved outcome to two years of very low-birthweight infants: Fact or artifact? *Developmental Medicine and Child Neurology, 28,* 579-588.

Koops, B.L., Morgan, L.J., & Battaglia, F.C. (1982). Neonatal mortality risk in relation to birth weight and gestational age: Update. *Journal of Pediatrics, 101,* 969-977.

Kravitz, H., Goldenburg, D., & Neyhus, A. (1978). Tactual exploration by normal infants. *Developmental Medicine and Child Neurology, 20,* 720-726.

Laszlo, J.I., & Bairstow, P. (1980). The measurement of kinaesthetic sensitivity in children and adults. *Developmental Medicine and Child Neurology, 22,* 454-464.

Law, M., Cadman, D., Rosenbaum, P., Walter, S., Russell, D., & DeMatteo, C. (1991). Neurodevelopmental therapy and upper extremity inhibitive casting for children with cerebral palsy. *Developmental Medicine and Child Neurology, 33*(5), 379-387.

Martin, G., & Pear, J. (1978). *Behavior modification: What it is and how to do it.* Englewood Cliffs, NJ: Prentice-Hall.

Moore, J. (1984). The neuroanatomy and pathology of cerebral palsy. *Selected proceedings from Barbro Salek memorial symposium* (pp. 3-58). Oak Park, IL: NDTA.

Michaels, J.M. (1990). Effects of physical therapy using a neurodevelopmental treatment approach on two children with spastic diplegia. Unpublished Master's thesis. Philadelphia: Hahnemann University.

Nashner, L.M., Shumway-Cook, A., & Marin, O. (1983). Stance posture control in select groups of children with cerebral palsy: Deficits in sensory organization and muscular coordination. *Experimental Brain Research, 49,* 393-409.

Neuro-Developmental Treatment Association, Inc. (1991). *Neuro-developmental treatment course syllabus.* Chicago, IL: Author.

Norkin, C., & Levange, P. (1983). *Joint structure and function, A comprehensive analysis.* Philadelphia: F.A. Davis.

Ophelia-Layman, P., Short, M.A., & Trombly, C. (1985). Kinesthetic recall of children with athetoid and spastic cerebral palsy and non-handicapped children. *American Journal of Occupational Therapy, 27,* 223-230.

Phelps, D.L., Brown, D.R., Tung, B., Cassady, G., McClead, R.E., Purohit, D.M., & Palmer, E.A. (1991). 28-day survival rates of 6676 neonates with birth weights of 1250 grams or less. *Pediatrics, 87,* 7-17.

Porter, R.E. (1987). Sensory considerations in handling techniques. In B.H. Connolly & P.C. Montgomery (Eds.), *Therapeutic exercise in developmental disabilities* (pp. 43-53). Chattanooga: Chattanooga Corporation.

Reeves, G. (1985). Influence of somatic activity on body scheme. *Sensory Integration Special Interest Section Newsletter, 8,* 1-2.

Riddle, D., Rothstein, J., & Lamb, R. (1987). Goniometric reliability in a clinical setting. *Physical Therapy, 67*(5), 668-673.

Rochat, P., & Senders, S.J. (1990). *Sitting and reaching in infancy.* Paper presented at the 7th International Conference on Infant Studies, Montreal.

Rochat, P. (1991). *Control of posture and action in infancy.* Paper presented at 11th Biennial Meetings of the International Society for the Study of Behavioral Development, Minneapolis.

Rochat, P. (in press). Self-sitting and reaching in 5-8 month-old infants: The impact of posture and its development on early eye-hand coordination. *Journal of Motor Behavior.*

Scherzer, A.L., & Tscharnuter, I. (1982). *Early diagnosis and therapy in cerebral palsy.* New York: Marcel Dekker.

Shapero Sabari, J. (1991). Motor learning concepts applied to activity-based intervention with adults with hemiplegia. *American Journal of Occupational Therapy, 45,* 523-530.

Shumway-Cook, A., & Horak, F.B. (1986). Assessing the influence of sensory interaction on balance: Suggestion from the field. *Physical Therapy, 66,* 1548-1550.

Stewart, A.L., Reynolds, E.O.R., & Lipscomb, A.P. (1981). Outcome for infants of very low birthweight: Survey of world literature. *Lancet, 1,* 1038-1041.

Umphred, C.A., & McCormack, G.L. (1985). Classification of common facilitatory and inhibitory treatment techniques. In D.A. Umphred (Ed.), *Neurological rehabilitation* (pp. 72-118). St. Louis, MO: Mosby.

Vogtle, L. (1984). The effect of shoulder and forearm rotation on hand function. In *Selected proceedings from Barbro Salek memorial symposium* (pp. 177-192). Oak Park, IL: NDT, Inc.

Valvano, J., & Long, T. (1991). Neurodevelopmental treatment: A review of the writings of the Bobaths. *Pediatric Physical Therapy, 3,* 125-129.

Wells, K. (1967). *Kinesiology: The scientific basis of human motion.* Philadelphia: W.B. Saunders.

Wilson, J.M. (1984). Cerebral palsy. In S.K. Campbell (Ed.), *Pediatric neurologic physical therapy* (pp. 353-413). New York: Churchill Livingstone.

# 8

# Upper-Extremity Casting: Adjunct Treatment for the Child with Cerebral Palsy

*Audrey Yasukawa*

## Introduction and Literature Review

Children with cerebral palsy (CP) have impairment of the central nervous system (CNS), which usually affects their ability to perform daily activities. The CNS lesion may produce the following problems in upper-extremity movement: (1) muscle incoordination of the normal patterns of arm and hand function, (2) poor voluntary control of contraction and relaxation of the spastic muscles, and (3) weakness of the muscle group opposing the spastic muscles. In selected cases a casting program may improve function of the arm or hand. Casting is a modality that directly impacts the problems identified in CNS lesions.

Few publications document the use of upper-extremity casting specifically for children with CP. Law, Cadman, Rosenbaum, Walter, Russell, and DeMatteo (1991) described the effect of neurodevelopmental treatment (NDT) and upper-extremity inhibitive casting, singularly or in combination, on hand function in 73 children with spastic CP. The children were stratified by severity and age, with the age range from 18 months to 8 years. They were randomized to one of four 6-month interventions: (1) intensive NDT plus cast, (2) intensive NDT, (3) regular NDT plus cast, and (4) regular NDT. Parent compliance and the age of the child had a significant effect on outcome. Parent active participation with the home program assisted the child's gain in hand function. The results suggested casting improved the quality of upper-extremity movement and range of motion. While the children who received intensive NDT and casting improved more in measures of hand function, the results were not statistically significant. Groen and Dommissee (1964) used plaster casting on 38 pediatric subjects with CP. Upper-extremity casting was applied to only three of the subjects. A long-arm cast was used extending from above the elbow to the fingertip. One of these children gained range of motion; the second child had poor results, and the third child withdrew from the study. The report failed to described the pre- and postcasting results, as well as the long-term results on functional control.

Single case reports have described the use of upper-extremity casting. Despite the limited generalization possible, a single case report can provide guidance and impetus for future research. Yasukawa (1990) reported the results of application of a long-arm cast to a child with hemiplegia. Casting and regular therapy were successful in improving arm and hand position and function. For the more severely involved child with spastic quadriplegia, upper-extremity casting has been used to prevent elbow flexion contractures and to assist caregivers with ease in handling the child (Cruickshank & O'Neill, 1990; Smith & Harris, 1985).

In rehabilitation centers, it is common practice for the occupational therapist to use upper-extremity casting to manage spasticity and contractures in individuals who have sustained a head injury (Booth, Doyle, & Montgomery, 1983; Hill, 1988). Casting often helps maintain upper-extremity range of motion and prevent contractures as the individual recovering from a head injury progresses through a period of spasticity. On the other hand, children with spastic cerebral palsy usually have a more stable course with respect to their neurological status. This may be one reason why casting the upper extremity in a child with CP has been controversial. In my clinical experience, the opinion in the medical profession is that casting gives no lasting functional results and that a more useful procedure with children of certain ages is surgery.

Casting has been a recommended adjunct treatment strategy for the lower extremity in children with CP (Cusick & Sussman, 1982; Watt, Sims, Harckhan, Schmidt, McMillan, & Hamilton, 1986). However, casting has not been a traditional modality for upper extremities of children with CP. In my clinical experience, it became frustrating to maintain gains in the upper-extremity range and function with these children. It appeared that the gains lasted for short periods and were difficult to carry over to the next treatment session. I began exploring the use of upper extremity casting as an adjunct to therapy to gain and maintain improved upper-extremity range and function. I was encouraged with the changes seen in selected children with spastic CP.

Upper-extremity involvement in children with CP presents clinical challenges to the therapist. Selection of a casting program requires a comprehensive analysis of the problem of CP and, more specifically, a careful study of the involved arm or hand. Each child must be carefully evaluated in order to determine whether or not casting would be beneficial. The selection of children is very important and requires clearly defined goals. The casting program must be individualized and carefully planned because of the extreme variations in how cerebral palsy affects the child's movement and ability to function.

## Common Upper-Extremity Problems in Spastic Cerebral Palsy

Spasticity creates major problems in upper-extremity motor development and function. Range of motion limitations of the elbow, forearm, wrist, and thumb are often targeted areas for change in the casting program. Proximal posturing, asymmetrical movement patterns, and contractures must always be considered in application of the cast. The common clinical presentation is described below. The classification system defined may be used to assist in communication among the medical team members.

The *shoulder* is generally positioned into internal rotation. There is decreased active reach above shoulder level. The initiation of active reach tends to be patterned with humeral abduction rather than humeral flexion and adduction.

The *elbow* is generally postured in some degree of flexion. At times, the active end range of full elbow extension is limited.

The internal rotation of the shoulder may affect the alignment distally. The *forearm* is generally positioned into pronation, which malaligns the wrist and hand for prehension and manipulation. Active forearm supination tends to be limited, more so when the elbow is fully extended than when flexed.

The *wrist* is often positioned in flexion and ulnar deviation. This position contributes to poor wrist stability and malalignment of the proximal joints and limits distal control. Some children with weak wrist extensors may compensate by using finger extensors.

Zancolli and Zancolli (1981) described a classification of deformities in the spastic hand according to the severity of flexion contracture of the wrist and fingers and paresis of the extensor group:

- *Group I.* These patients can extend the fingers when the wrist is positioned between neutral and 20° of flexion. Flexion spasticity is minimal.
- *Group II.* These patients can extend the fingers when the wrist is flexed more than 20°. This group has been further subdivided according to severity of spasticity and the control of the extensor muscles of the wrist.
    - 2a. The patient can actively extend the wrist with fingers flexed. The wrist extensors are active and the principal spasticity is in the finger flexor muscles.
    - 2b. The patient cannot extend the wrist with fingers flexed. The wrist extensor is either of poor-to-trace strength or totally absent.

The influence of the forearm and wrist will affect position of the *fingers*. An intrinsic plus or swan-neck deformity may occur secondary to the muscle imbalance and ligamentous or capsular relaxation. A flexion or extension deformity may occur depending on the stronger muscle group pull.

The *thumb* may be positioned in the palm, tightly adducted, or flexed. Instability of the metacarpophalangeal (MCP) joint secondary to overstretched ligaments and tendons may occur, which leads to hypermobility and pain.

House, Gwathmey, and Fidler (1981) described a classification of thumb deformities based on the recognition that the position of the forearm and wrist are intimately related to the function of the thumb.

- *Type I.* The metacarpal is held in adduction and influenced by a combination of spasticity and fixed contracture in the adductor pollicis and first dorsal interosseus.
- *Type II.* The metacarpal is held in adduction and MCP in flexion. In addition to the description in Type 1, the MCP joint is influenced by spasticity in and/or shortened length of the flexor pollicis brevis muscle.
- *Type III.* The metacarpal is held in adduction in combination with the MCP joint in hyperextension. The MCP joint of the thumb is very unstable secondary to the poor alignment of the

joint, causing overstretching during active extension.

- *Type IV.* The metacarpal is held in adduction in combination with flexion deformities in the MCP and interphalangeal joints. This is considered the most severe deformity; it may be influenced by the spasticity and shortened length in the flexor, adductor, and intrinsic muscles.

To summarize, candidates for casting often exhibit upper-extremity weakness that is characterized by muscle imbalance due to spasticity of specific muscle groups and weakness of the opposing muscle groups. This imbalance causes poor alignment and stability of the joints, which limits the child's ability to isolate controlled fine movements.

## General Factors to Consider for Casting

The goals of upper-extremity casting are similar to those of upper-extremity surgery in children with spastic CP. Communication with the child's physician is necessary prior to initiating a casting program. Prerequisites established for successful surgical candidates may be applied to casting candidates (House, Gwathmey, & Fidler, 1981; Keats, 1965; Mital & Sakellarides, 1981; Samilson, 1966). The major goal is improvement in arm and hand function. General factors to be considered before casting include:

1. *Motivation of the child.* The child must exhibit a willingness and ability to cooperate and follow the retraining program. The child must exhibit some motivation to use his or her upper extremities to take advantage of the functional improvement from casting.
2. *Behavioral control.* The child should exhibit no significant behavioral or cognitive problems that may interfere with his or her ability to cooperate during the casting program.
3. *Sensation.* Most children with CP are believed to have some impairment of stereognosis and proprioception (Goldner & Ferlic, 1966; Tizard, Pain, & Crothers, 1954). Severe sensory impairment may be a limiting factor in the functional motor outcome to be achieved through casting.
4. *Type and degree of involvement.* The child with spastic hemiplegia may achieve more improvement in hand function than a child with spastic quadriplegia. A child with mild involvement may improve more than a child who is severely involved. Other factors may also limit functional outcomes of upper-extremity casting. These include significantly decreased head and trunk control, retained primitive reflex patterns influencing head position and upper extremities, compensatory fixing patterns, and severe involvement of the lower extremities.
5. *Voluntary control.* The amount of voluntary motor control at the joints to be casted is an important indicator for casting. The child must be able to activate individual muscles or muscle groups on command. The extent of voluntary control indicates the potential for active use if a more functional balance of the muscles can be achieved. For example, a child with mild impairment may be a good candidate if good control of the opposing muscle groups is present.
6. *Informed parents.* Parents must be aware that the involved upper extremity will never demonstrate "normal" function. They should have a clear understanding of the goals of the casting program. Parents should also be informed of any time commitment necessary to effectively implement a casting program, including follow-up activities at home. Once fully informed, parents should be allowed to decide with the therapist and physician the best course of action.
7. *Age of child.* The optimal age when a casting program should be implemented must be determined on an individual basis. In my clinical experience, casting may be successfully initiated as early as 15 months of age or when the child is beginning to creep or pull to a standing position. Early use of casting allows the child to spontaneously adapt to the casting process while the muscles, tendons, and joints are pliable.

It is difficult to cast a child under 3 years of age and expect follow-up on a specific exercise program. Similar to the ongoing therapy sessions, casting may be a long-term process. A child may only tolerate 2 weeks of wearing a cast and may not comply with wearing an arm splint. Range of motion will not be maintained without compliance to the wearing schedule. Casting can be reinstated at any time during the child's therapy program. We have found that most parents prefer not having their child casted during the hot summer months or over special holiday seasons.

## Assessment

A baseline assessment is important to determine an appropriate casting program and to compare function of the involved arm, pre- and postcasting. An effective casting program can be developed only after a detailed assessment of the child's range of motion, movement patterns, sensation, and specific functional needs.

Normal arm and hand movements are characterized by a vast range of skilled purposeful actions. Upper-extremity movement requires a delicate balance of stability for arm placement and to hold an object, as well as mobility for precision grasp, release, and manipulation. For a child with cerebral palsy, upper-extremity movement problems are complex due to the muscle imbalance that exists between the extrinsic and intrinsic muscles, increased muscle tone, muscle weakness, and/or contractures. Furthermore, function of the arm and hand are so interdependent that any malalignment in one joint will affect other related joints.

The therapist considers the interdependence of upper-extremity movements in evaluating arm function. In the child with cerebral palsy, the combination of hypertonic agonist muscles and weaker antagonist muscles results in muscle imbalance with the more spastic muscle action predominating. A primary objective of therapy is to improve muscular balance between the agonist and antagonist; therefore, it is essential to identify the action of specific muscles in the motor patterns observed.

Before considering casting the distal portion of the upper extremity (e.g., hand and wrist), the therapist should evaluate the proximal function of the shoulder, elbow, and forearm to determine if it is adequate to position the hand in space. If these prerequisites are not met, it may be necessary to treat these problems first. For example, forearm supination assists with alignment and permits extension of the wrist and fingers out of the palm. Therefore, casting of the forearm in supination may be needed before casting the wrist to facilitate prehension.

Orthopedic problems may prohibit the use of casting. Communication with the child's physician and use of radiographs may be necessary to rule out problems that contraindicate the use of casting.

## Indications for Casting and Goal Setting

There are many variables to consider when selecting appropriate children for casting. The decisions regarding when and for how long to cast will depend on the clinical judgment and experience of the therapist, as well as the families' priorities and involvement. The aim of the casting program may include the following: (1) to correct deformity or shortened muscles, (2) to restore muscle balance by reducing the pull of spastic or shortened muscles, (3) to strengthen the opposing muscle group, (4) to improve joint stability, (5) to secure a functional grasp and release mechanism, and (6) to provide a pinch or improve existing pinch.

If the child has poor voluntary control and flaccid or weak muscles among the spastic muscles, casting should be directed toward maintaining range of motion and hygiene care. It is contraindicated to cast a child with excessive hypermobility of the joint, athetosis, and/or severe sensory disorder (Keats, 1965; McCue, Honner, & Chapman, 1970; Skoff & Woodbury, 1985).

After the therapist has carefully evaluated the child's upper-extremity function, the therapist and parents must decide whether a casting program and a specific type of cast would achieve the end goal of improving upper-extremity function, comfort, or ease in caring for the child. The muscle imbalance between spastic and weaker muscles should be kept in mind. The best results can only be achieved by careful selection of the child, along with intensive therapy during and after the casting program.

The short-term goals of the casting program should be small, qualitative changes of the shoulder or joints being casted. One problem often encountered by therapists when casting the upper extremities in children with CP is establishing goals that are too ambitious, with expectations of restoring normal function.

Casting a child with contractures and/or poor voluntary control over motor function should be directed toward maintaining range of motion or preventing further contracture. It is not appropriate to establish a goal to improve fine-motor control. Another goal in casting can focus on improving the ease with which the caregiver dresses or bathes the child. When discussing the goals of casting with the family, the therapist should inform the parents that the spasticity can recur. When discussing goals, it should be stressed that the child may improve in function or remain unchanged, depending on available motor control and family involvement.

## Casting Procedures

Prior to initiating a casting program, the therapist should discuss the following with the parent and/or caregiver: (1) the goals of the casting program, (2) the importance of follow-up and assisting the child with an active home program, and (3) regular weekly attendance of occupational therapy.

Therapists must employ proper casting techniques and procedures, including careful monitoring of cast fit, to prevent skin breakdown. The therapists casting and the individual holding the child's extremity during casting must work together as a team.

The cast is initially applied at a submaximal range, generally about 5 to 10 degrees less than the range available with maximal stretch. The casts are applied in a series and changed every 3 to 7 days, depending on the type of cast. The casts are marked with the date of application and projected date of removal. Parents are instructed to monitor the cast daily by checking the skin proximal and distal to the cast, and are provided with written precautions that would indicate the need for immediate cast removal. They are also instructed in emergency procedures should the child need immediate cast removal.

When changing the cast, the therapist checks for

potential skin problems and cleans the casted area. The upper-extremity movement and range are reassessed and a new cast immediately reapplied. We recommend that a maximum of five to six casts be applied in the series. The casting program may be terminated when clinical improvements are noted in the following areas: correction of deformity, restoration of muscle balance, improvement in motor skills, and/or attainment of precasting goals. At the end of the casting program, the final cast may be bivalved into anterior and posterior shells and used to maintain the gains from the serial casting program. The therapist may prefer to use a sturdier material, such as fiberglass or a lightweight contoured orthotic material, as the follow-up bivalve splint.

After the casting program, an aggressive therapy and home program must then be initiated to maintain the new muscle tendon length and to strengthen the opposing muscle groups.

## Types of Casts

After the initial assessment, the therapist selects the type of cast that may assist with improving function. The basic types of casts used with this population will be described with considerations for their use. Other variations of casts will not be included in this section but are described in the literature (Feldman, 1990; Yasukawa & Hill, 1988). The therapist must use clinical judgment and creative problem solving to make the final decision about which cast to select.

### Drop-Out Cast (Humerus Enclosed)

The drop-out cast with the humerus enclosed is used to increase elbow range quickly (Figure 1). The forearm shell prevents flexion of the arm but allows the elbow to extend. This is most effective for a child with severe elbow flexion contracture, greater than 45° elbow flexion. When the range of elbow extension increases 10° to 15°, a new cast should be applied. A rigid circular elbow cast is preferable if the range in elbow extension is less than 45°

### Rigid Circular Elbow Cast

A rigid circular elbow cast is used to gradually increase elbow range (Figure 2). This cast is left in place for 5 to 7 days to provide a slow, gradual stretch. This cast can be quickly applied and provides equalized pressure. Due to the ease of application, it may be used on upper extremities with severe spasticity. However, it is generally used on an elbow with minimal to moderate flexion contracture.

### Long-Arm Cast

The long-arm cast incorporates the elbow, forearm, and wrist; the thumb may or may not be included (Figure 3). This cast can gradually increase forearm

**Figure 1. Drop-Out Cast (Humerus Enclosed)**

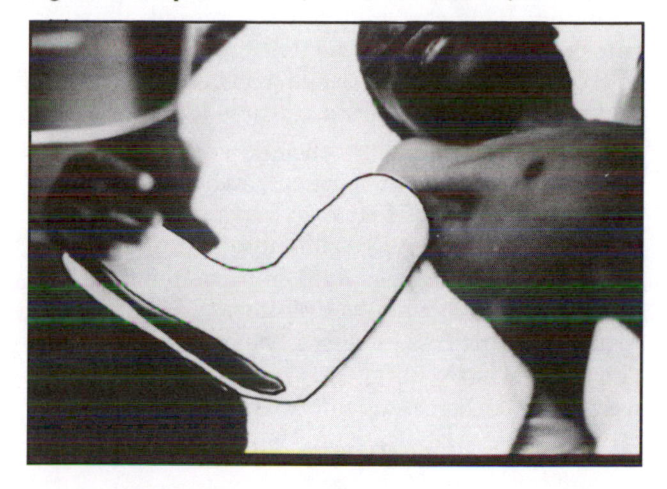

**Figure 2. Rigid Circular Elbow Cast**

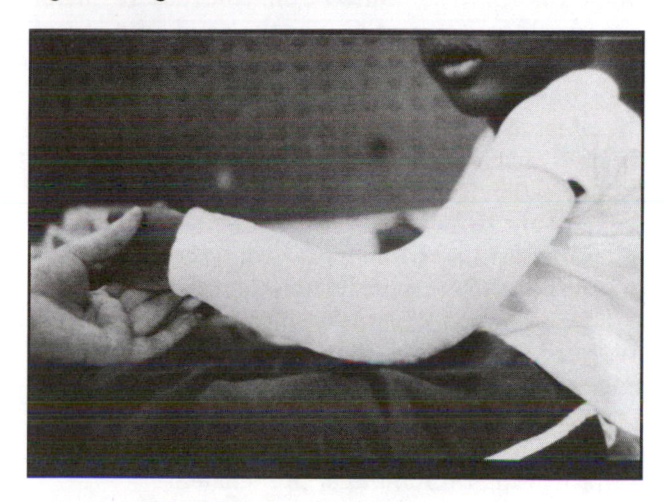

**Figure 3. Long-Arm Cast**

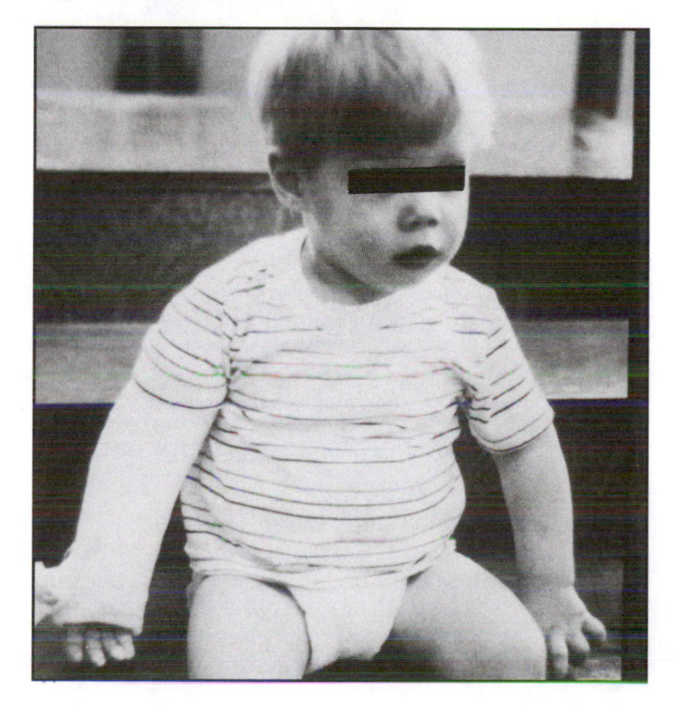

rotation and assist with lengthening the opposing pronators. This cast may not be appropriate when extreme tightness is present throughout the arm. Microtearing of the soft tissue may occur when tight muscles are stretched over multiple joints.

The long-arm cast has been an excellent cast to use on the child with minimal involvement in the hemiplegic arm. The therapist may find that the cast can serve as an extra pair of hands during a handling session. With the cast in place, improvement in proximal stability of the shoulder girdle may occur through regular therapy session and a home program. As the forearm and elbow are gradually lengthened, improvements may be seen in active forearm supination control.

## Wrist Cast

The wrist cast is used to gradually increase wrist range that may be limited from contracture and/or abnormal tone (Figure 4). For the child with severe contracture and decreased muscle control, this cast may improve the alignment of the hand allowing the therapist to fabricate a follow-up maintenance splint for skin and hygiene care. For the child who demonstrates difficulty in differentiating wrist and finger movements, this cast can provide wrist stability and allow better distal fine-motor control. This cast is appropriate when there is muscle imbalance of the wrist and hand. Weakness may be present in the intrinsic and/or extrinsic musculature. The wrist and fingers are influenced by spasticity that inhibits grasp, release, and manipulation of objects. This cast can stabilize the wrist and provide wrist alignment for promoting thumb-finger prehension.

## Wrist Cast with Thumb Enclosed

The wrist cast with thumb enclosed is used for the thumb-in-palm deformity (Figure 5). This cast can gradually increase the web space and palmar expansion. By elongating the thumb flexor and adductor muscles, this cast may promote the rebalancing of the thumb extensor and abductor muscles. Positioning the thumb out of the palm gradually has an inhibiting effect on the spasticity of the finger and wrist flexors. This type of cast may be contraindicated if the thumb metacarpophalangeal joint is subluxed or hypermobile.

## Platform Cast

The platform cast is used to gradually improve wrist and finger range (Figure 6). The child generally exhibits isolated muscle control of the hand but continues to be influenced by mild flexor tone in the extrinsic musculature, the intrinsics, or both. When the child attempts to extend the fingers, slight flexion of the long finger flexors, intrinsic tightness, and/or thumb flexion are evident. The platform cast can be applied with or without the thumb enclosed. As in

**Figure 4. Wrist Cast**

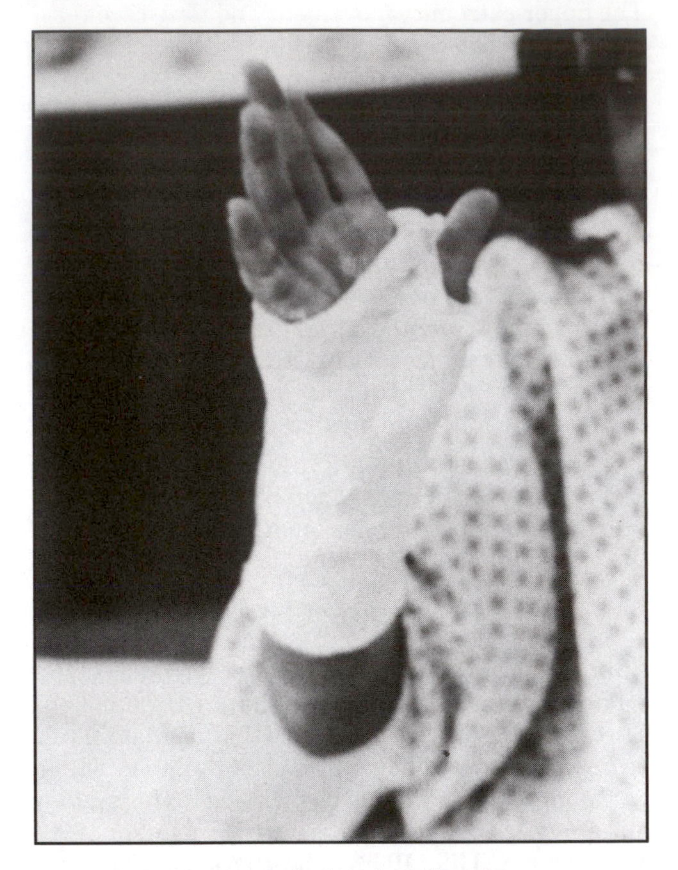

**Figure 5. Wrist Cast with Thumb Enclosed**

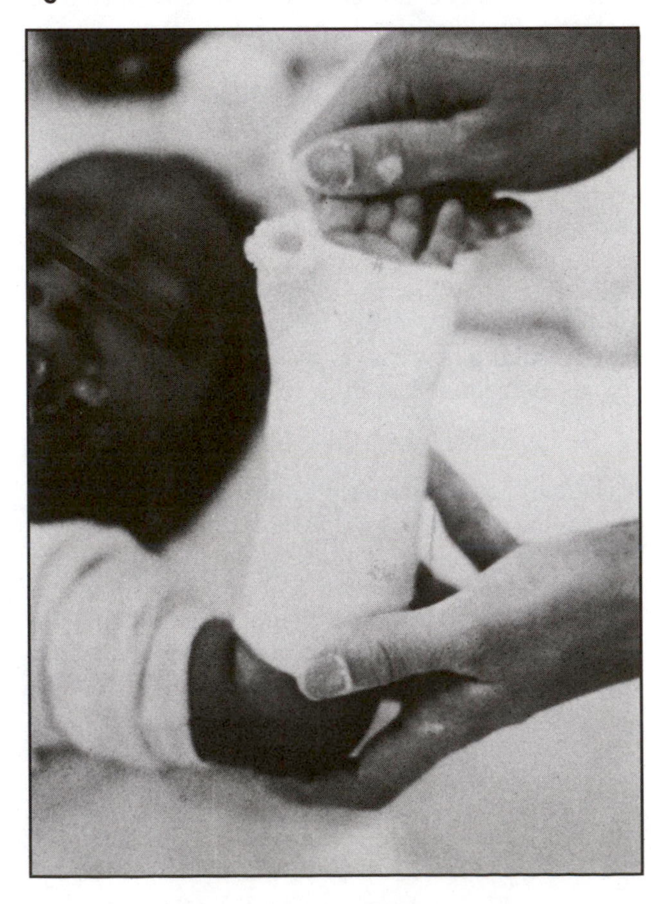

**Figure 6. Platform Cast**

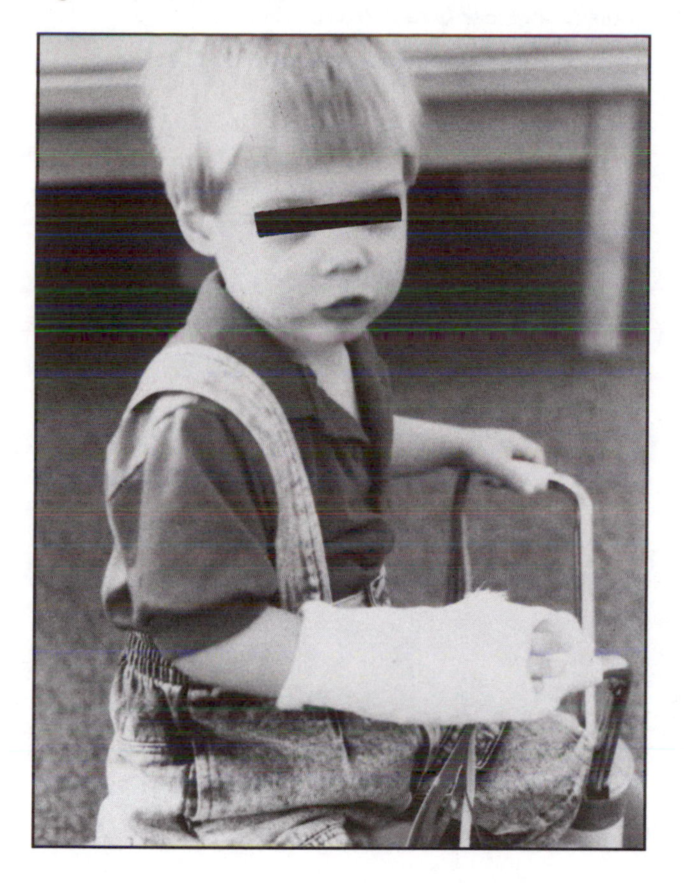

fabricating a resting hand splint, the fingers must be carefully positioned in the platform portion of the cast. Pressure must be avoided at the distal interphalangeal joints to prevent deformity of the proximal and distal interphalangeal joints. This cast can gradually relax tightness of the hand and act as a weight-bearing cast. It can position the wrist and fingers into gradual extension to facilitate an active grasp and release mechanism.

## Follow-Up

Upper extremity casting is an adjunct to therapy. Parental participation and follow-up in the casting program are essential to a successful outcome. The occupational therapist's follow-up includes monitoring range of motion and implementing a program to restore the balance of the arm's muscle actions. Therapy and home programs focus on active exercise. Phelps (1957) evaluated the long-term results of lower-extremity surgical procedures for children with cerebral palsy. He ascribed poor maintenance of gains to the lack of follow-up therapy. He attributed surgical failures to inconsistent use of the night brace and a lack of sufficient motor power to maintain the correction. He stated that children who did well and maintained benefits of the surgical procedures to the lower extremities were those who wore a night brace throughout the period of growth. In keeping with Phelps's recommendation, continued use of a bivalve cast may be needed after a casting program to maintain the newly acquired muscle length.

Each child will respond differently to the casting program. The follow-up measure of wearing a bivalve cast will depend upon individual needs. The primary goal of the bivalve cast is to maintain the length of the spastic muscle and range of motion. The therapy sessions and home program will focus on improving strength of the opposing muscle group. The therapist must continue to reassess the child, focusing on gradually decreasing the wearing schedule as the child is able to maintain control without wearing the bivalve cast.

The therapist may prefer using fiberglass or orthotic materials for follow-up management. It is important to carefully evaluate and monitor the use of the orthotic device of choice. The orthosis or bivalve cast should maintain range, provide equalized pressure, and provide fit throughout the arm.

For the severely involved upper extremity with contractures and stiffness, a maintenance program of positioning and range of motion is important factor to consider prior to casting and throughout the casting program. If the attempts of a casting program and standard therapeutic techniques have failed, surgery may be warranted to assist with improving hygiene and/or function of the upper extremity. At times, a casting program has been used to assist the surgeon in deciding on the type of surgical procedure a child may benefit from to improve function.

## Summary

Many variables need to be considered in evaluating the effect of upper-extremity casting of children with cerebral palsy. Casting does not follow a set protocol; on the contrary, each child has unique needs and will have a unique casting program. The therapist monitors the child's response during the casting program and adjusts the treatment accordingly. The decision to cast depends a great deal on the assessment of the whole child and parental involvement. The child with mild deformity and muscle imbalance with potentially good muscle control can be expected to demonstrate significant improvements with casting and follow-up. The child with a nonfunctional arm may demonstrate an improved position and will require a maintenance program. The benefits achieved by the child in between these ranges is determined by the child's cognitive, psychosocial, and motor function as well as the parents' involvement.

The therapist is responsible for determining whether or not a child would benefit from casting and designing a program based on the child's individual needs and the family's priorities. We must recognize that children and their families differ in their response to a casting program.

In casting children with spastic cerebral palsy, it is important to follow up and investigate the reasons as to why the casting program was successful or unsuccessful in order to identify the candidates who are most likely to have positive results. Hopefully, this information can serve as a base from which therapists can begin to solve the complex problems encountered. More research is needed to confirm the theoretical foundations for casting and to determine if children with cerebral palsy may benefit from this type of intervention. Upper-extremity casting for this population provides challenge and opportunities for occupational therapists. Since this modality is relatively new in the management of children with CP, therapists must cautiously forge ahead toward the development of new, more effective casting procedures.

## Case Studies

The following case studies are examples of applying the basic principles that were discussed in the content of the chapter.

### Case 1

Mark was initially seen for evaluation of possible casting at 6 1/2 years of age. He was diagnosed as having CP with spastic quadriplegia and mental retardation. He was seen at another facility for weekly occupational therapy. The major limiting factor with his progress in therapy was poor therapy attendance and limited follow-up at home. Mark lived with his father who was his primary caregiver.

In the initial assessment, Mark demonstrated characteristics typical of spastic quadriplegia with more involvement on his right side. He had poor trunk control with an extremely rounded thoracic and lumbar spine and shortening of the neck extensors. His lower extremities pulled into a crossed extensor pattern with no dissociation and obvious muscle tightness. Mark was able to roll using an atypical pattern but could not assume any higher level positions against gravity. His mobility was limited secondary to his severe right upper-extremity contractures. His right elbow and wrist were positioned into flexion; the forearm was pronated and the shoulder girdle was protracted (Figure 7). Mark was influenced by the increased abnormal tone and moved in a flexor synergy pattern. He preferred to use his left upper extremity for simple functional fine-motor activities and used an immature palmar grasp for writing.

The physician, therapist, and father decided to initiate an intensive serial casting program to increase range of motion and to prevent skin breakdown to the right arm. The main objective was to improve ease of dressing and bathing the child. The initial assessment of the right upper-extremity passive range of motion revealed the following: The elbow lacked 90° of extension, wrist flexion was 0° to 120°, wrist extension was fixed at neutral, the forearm was postured in pronation, and supination was at 0°.

**Figure 7. Elbow and Wrist in Flexion; Forearm Pronated; Shoulder Girdle Protracted**

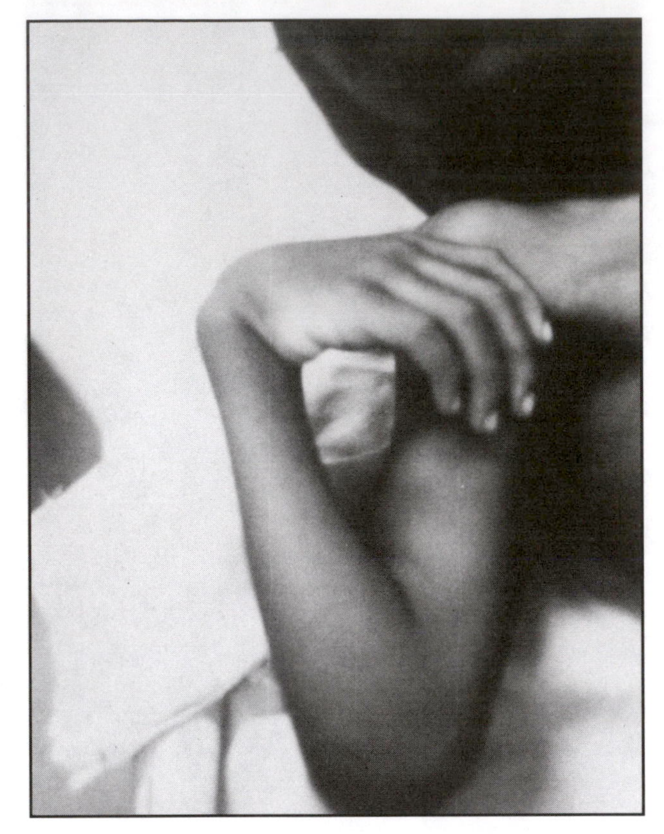

The child was scheduled for weekly occupational therapy sessions during the casting program. Since the arm was severely contracted, the therapist initially applied a drop-out cast to improve range in the elbow. After a week, the cast was removed, skin checked and cleansed, and elbow range was measured. Mark gained 15° of elbow extension and a rigid circular elbow cast was immediately applied (Figure 8). For the next 4 weeks, Mark continued to gain 10° to 15° increase of elbow extension. After the 6th week of casting the elbow, he lacked 15° of elbow extension with maximal stretch (Figure 9). The resting position of his elbow was lacking 55° of elbow extension.

The wrist was then casted with the first cast positioned at neutral (Figure 10). After 3 weeks of casting his wrist, he attained full range into wrist extension (Figure 11). A final long-arm cast was applied to gradually improve elbow extension with forearm supination (Figure 12). The end ranges from the 9 weeks of intensive casting were as follows:

| | Precasting | Postcasting |
|---|---|---|
| Elbow extension | 155°–90° | 155°–35° |
| Wrist extension | 0° | 0°–80° |
| Forearm | 0° | 0°–45° |

Figure 8. 15° Elbow Extension; Rigid Circular Elbow Cast

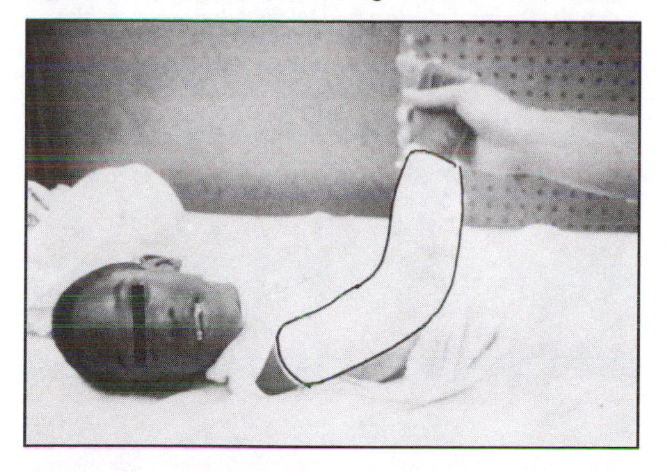

Figure 9. Sixth Week of Elbow Casting

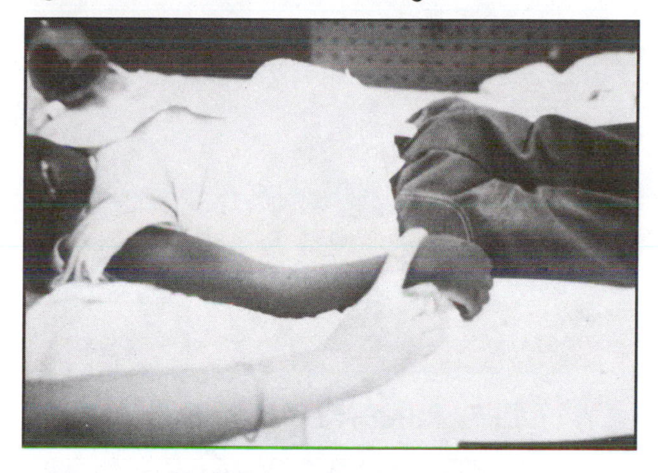

Figure 10. Wrist Casted in Neutral Position

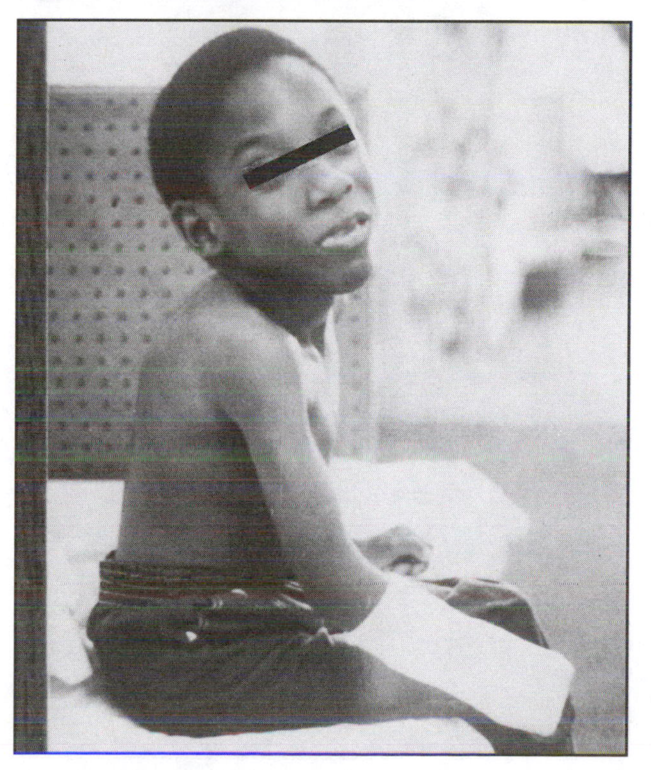

Figure 11. After 3 Weeks of Casting, Wrist is in Full Extension

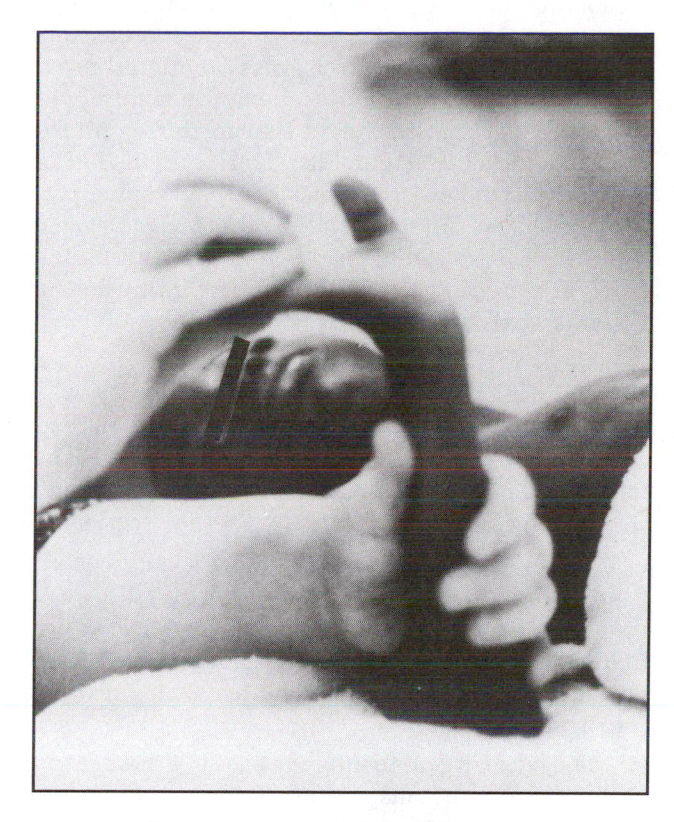

Figure 12. Final Long-Arm Cast is Applied to Improve Elbow Extension with Forearm Supination

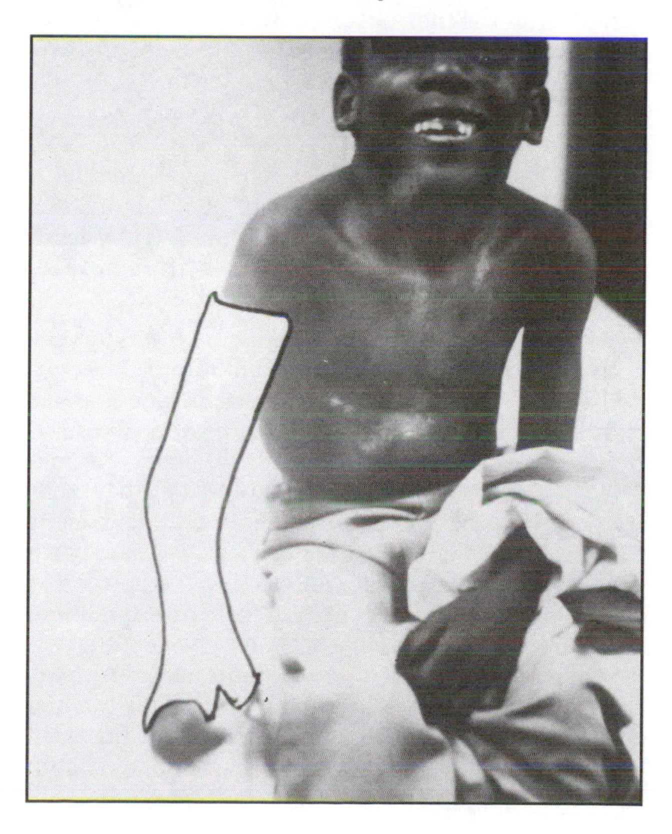

One surprising result that occurred with the gradual lengthening into elbow extension was activation of the triceps. The occupational therapy program focused on the following: to increase active use of the right arm as a gross functional assist for tabletop activities and to have the right upper extremity assist with weight bearing during transitional movements (Figure 13). Mark and his father were instructed on an intensive strengthening program at home. After casting, it was recommended that Mark wear a long-arm splint to maintain the newly acquired length until he could maintain control without wearing the splint.

Mark was followed in the outpatient clinic for reevaluations. The right upper extremity was nonfunctional and the flexion posture returned (Figure 14). The father had extreme difficulty with the follow-up home program and was unable to attend the ongoing weekly occupational therapy sessions.

A child with spastic quadriplegia may benefit from a casting program, but it is important to keep in mind the general factors that influence successful maintenance of gains achieved during casting. Parent education and participation are some of these important factors.

The father's goal to improve ease of dressing and bathing the child may have "gotten lost" as the therapist discovered more range and active use of the right upper extremity were possible. The father's understanding of the total care and ongoing responsibility may have become unclear. Therefore, it is essential for the therapist and the child's caregiver to collaborate and communicate the purpose of the casting and follow-up maintenance program.

## Case 2

Tammy was 5 1/2 years of age when evaluated for a casting program. At 8 months of age, her mother noticed she was not using her right hand. A physician confirmed Tammy's diagnosis of CP with right spastic hemiplegia at 10 months of age.

In the initial assessment, Tammy was a very active and cooperative child. She ambulated independently with her right foot in plantar flexion and her right arm and hand in flexion. Tammy actively raised her right arm full range overhead, although asymmetry was noted when comparing shoulder girdle stability with the uninvolved arm. There was also slight flexion of the elbow during overhead reach (Figure 15). Tammy isolated active flexion and extension of the elbow but had poor control of forearm supination. She lacked wrist stability and had poor differentiation of wrist and hand function. Tammy used a gross grasp and release pattern and did not have the fine thumb-fingertip prehension (Figure 16). An increased resistance to movement was noted at the end range of forearm supination and wrist extension.

**Figure 13. Attempt to Increase Active Use of Right Upper Extremity During Tabletop and Weight-Bearing Activities**

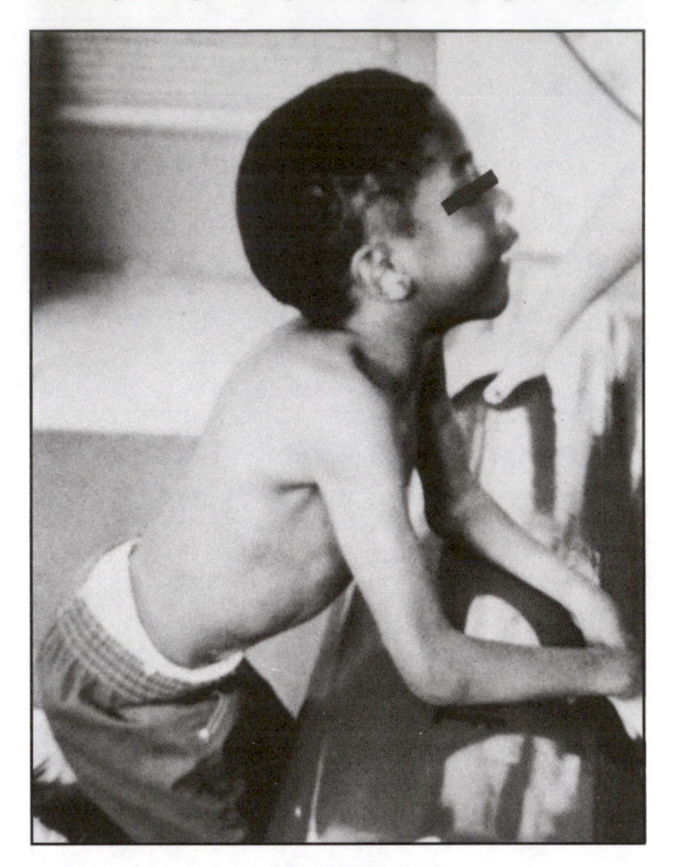

**Figure 14. Return of Flexion Posture**

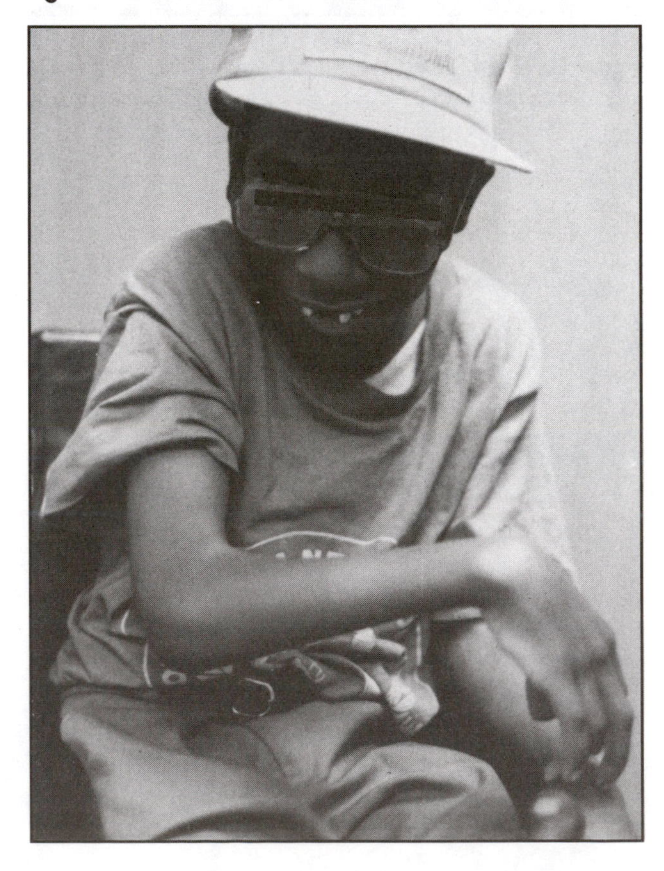

Figure 15. Slight Flexion of Elbow During Overhead Reach

Figure 16. Gross Grasp and Release Pattern; No Fine Thumb–Fingertip Prehension

Figure 17. Long-Arm Cast

Figure 18. Gaining Control of Forearm Supination with Elbow Flexed

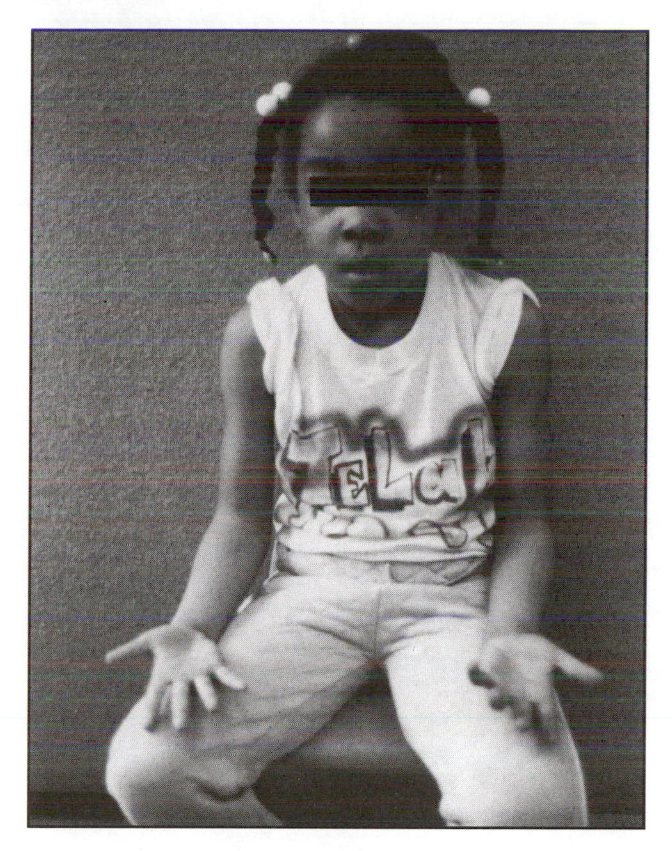

During play and physically stressful tasks, Tammy demonstrated associated reactions with increased spasticity in the affected side. Tammy demonstrated a spontaneous protective response with the right arm, although delayed. Generally, she used her unaffected hand for initiating reach and grasp and for manipulating objects. She used her involved hand as an assist for holding objects.

A long-arm cast was applied to: (1) increase shoulder girdle stability and strength and (2) increase active forearm supination with elbow flexed and extended. The casting program continued for 3 weeks, with casts removed weekly and new ones applied to accommodate the increased range. While wearing the cast, treatment was aimed at strengthening the shoulder girdle and active overhead reach.

After the casting program, a bivalved long-arm cast was fabricated to maintain the newly acquired length of elbow extension with forearm supination. Tammy wore this bivalve cast every night and continued with her outpatient occupational therapy sessions. Her home program consisted of strengthening the opposing muscle groups.

As Tammy developed greater control of forearm supination, a wrist cast was applied to improve grasp and facilitate thumb to fingertip prehension. The

**Figure 19. Gaining Control of Forearm Supinatin with Elbow Extended**

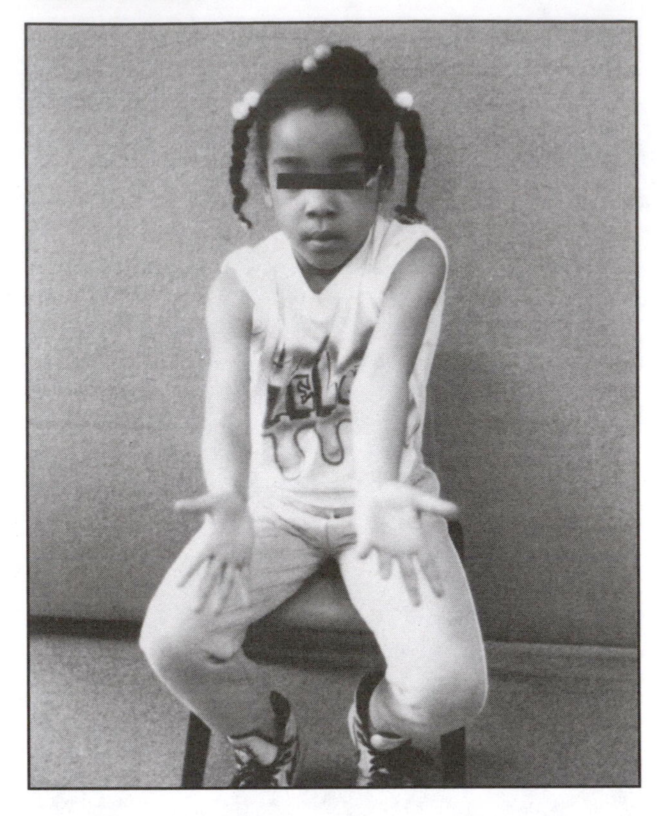

**Figure 20. Spontaneous Use of Impaired Hand to Assist During Bimanual Activities**

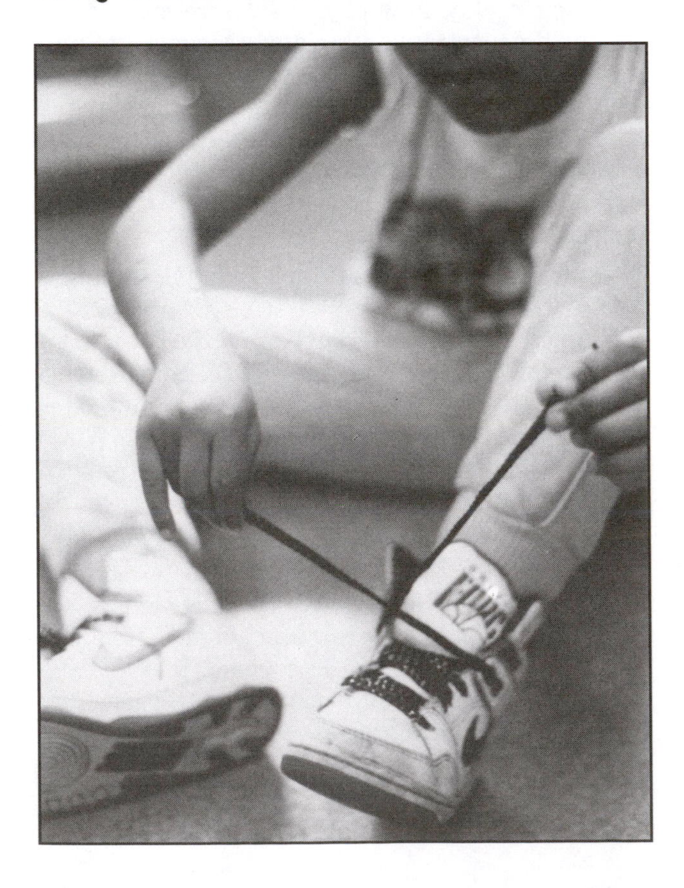

wrist was casted for 3 weeks and changed weekly. Her therapy sessions and home program consisted of activities requiring grasp and release of various sized objects and incorporating forearm supination.

The immediate result of the casting program was overall improvement in Tammy's strength and control of her hemiplegic arm. The gradual lengthening of the shortened muscle group with the long-arm cast encouraged Tammy to actively reach overhead with increased stability of the shoulder girdle and full elbow extension (Figure 17). The balance of tone in the forearm allowed Tammy to actively supinate. Through a strengthening program, Tammy gained control of forearm supination with her elbow flexed and extended (Figures 18 and 19).

As she developed forearm stability and control, she continued to improve her finer prehension skills with a wrist cast. She grasped pegs using a superior pincer and spontaneously used the impaired hand to assist during bimanual activities (Figure 20).

Tammy maintained the gains noted from the casting program and continued to improve with weekly outpatient occupational therapy. The new strength and control of the right upper extremity needed continual monitoring and reassessment to refine and encourage greater right upper-extremity skills. It was anticipated that her gains would be maintained with consistent follow-through and by active participation on the part of the caregiver.


## Acknowledgment

I would like to thank the rehabilitation staff and physicians at La Rabida Children's Hospital for their support. I would also like to thank the families and children who tried my ideas, gave me feedback, and offered me so many helpful suggestions and solutions.


## References


Booth, B.J., Doyle, M., & Montgomery, J. (1983). Serial casting for the management of spasticity in the head injured adult. *Physical Therapy, 63,*(12), 1960-1966.

Cruickshank, D.A., & O'Neill, D.L. (1990). Upper extremity inhibitive casting in a boy with spastic quadriplegia. *American Journal of Occupational Therapy, 44* (6), 552-555.

Cusick, B., & Sussman, M. (1982). Short leg casts: Their role in the management of cerebral palsy. *Physical and Occupational Therapy in Pediatrics, 2*(3/4), 93-110.

Feldman, P.A. (1990). Upper extremity casting and splinting. In M.D. Glenn & J. Whyte (Eds.), *The practical management of spasticity in children and adults.* Malvern, PA: Lea & Febiger.

Goldner, J.L., & Ferlic, D.C. (1966). Sensory status of the hand as related to reconstructive surgery of the upper extremity in cerebral palsy. *Clinical Orthopedics, 46,* 87-92.

Groen, J.A., & Dommissee, G.R. (1964). Plaster casts in the conservative treatment of cerebral palsy. *South Africa Medical Journal,* 502-505.

Hill, J. (1988). Management of abnormal tone through casting and orthotics. In K. Kovich & D. Bermann (Eds.), *Head injury: a guide to functional outcomes in occupational therapy.* Gaithersburg, MD: Aspen Publishers.

House, J.H., Gwathmey, F.W., & Fidler, M.O. (1981). A dynamic approach to the thumb-in-palm deformity in cerebral palsy. *Journal of Bone and Joint Surgery, 63-A*(2), 216-225.

Keats, S. (1965). Surgical treatment of the hand in cerebral palsy: Correction of thumb-in-palm and other deformities. *Journal of Bone and Joint Surgery, 47-A*(2), 274-284.

Law, M., Cadman, D., Rosenbaum, P., Walter, S., Russell, D., & DeMatteo, C. (1991). Neurodevelopmental therapy and upper extremity inhibitive casting for children with cerebral palsy. *Developmental Medicine and Child Neurology, 33,* 379-387.

McCue, F.C., Honner, R., & Chapman, W.C. (1970). Transfer of the brachioradialis for hands deformed by cerebral palsy. *Journal of Bone and Joint Surgery, 53A*(6), 1171-1181.

Mital, M.A., & Sakellarides, H.T. (1981). Surgery of the upper extremity in the retarded individual with spastic cerebral palsy. *Orthopedic Clinics of North America, 12*(1), 127-141.

Phelps, W.M. (1957). Long-term results of orthopedic surgery in cerebral palsy. *Journal of Bone and Joint Surgery, 39-A*(1), 53-59.

Samilson, R.J. (1966). Principles of assessment of the upper limb in cerebral palsy. *Clinical Orthopedics and Related Research, 47,* 105-115.

Skoff, H., & Woodbury, D.F. (1985). Current concepts review-management of the upper extremity in cerebral palsy. *Journal of Bone and Joint Surgery, 67-A*(3), 500-503.

Smith, L.H., & Harris, S.R. (1985). Upper extremity inhibitive casting for a child with cerebral palsy. *Physical and Occupational Therapy in Pediatrics, 5*(1), 71-79.

Tizard, J.P.M., Pain, R., & Crothers, B. (1954). Disturbances of sensation in children with hemiplegia. *Journal of the American Medical Association, 55,* 628-632.

Watt, J., Sims, D., Harckhan, F., Schmidt, L., McMillan, A., & Hamilton, J. (1986). A prospective study of inhibitive casting as an adjunct to physiotherapy for cerebral palsied children. *Developmental Medicine and Child Neurology, 28,* 480-488.

Yasukawa, A. (1990). Case report—upper extremity casting: adjunct treatment for a child with cerebral palsy hemiplegia. *American Journal of Occupational Therapy, 44*(9), 840-846.

Yasukawa, A., & Hill, J. (1988). Casting to improve upper extremity function. In R. Boehme (Ed.), *Improving upper body control—An approach to assessment and treatment of tonal dysfunction.* Tucson, AZ: Therapy Skills Builders.

Zancolli, E.A., & Zancolli, E.R. (1981). Surgical management of the hemiplegic spastic hand in cerebral palsy, *Surgery Clinics of North America, 61*(2), 395-406.